THE ROMAN ART OF WAR

In memory of my father,
Brian Ellis Gilliver

The Roman Art of War

C.M. GILLIVER

TEMPUS

First published 1999
First paperback edition 2001

PUBLISHED IN THE UNITED KINGDOM BY:

Tempus Publishing Ltd
The Mill, Brimscombe Port
Stroud, Gloucestershire GL5 2QG

PUBLISHED IN THE UNITED STATES OF AMERICA BY:

Tempus Publishing Inc.
2A Cumberland Street
Charleston, SC 29401

Tempus books are available in France, Germany and Belgium
from the following addresses:

Tempus Publishing Group
21 Avenue de la République
37300 Joué-lès-Tours
FRANCE

Tempus Publishing Group
Gustav-Adolf-Straße 3
99084 Erfurt
GERMANY

Tempus Publishing Group
Place de L'Alma 4/5
1200 Brussels
BELGIUM

British Library Cataloguing in Publication Data.
A catalogue record for this book is available from the British Library.

ISBN 0 7524 1939 0

Typesetting and origination by Tempus Publishing.
PRINTED AND BOUND IN GREAT BRITAIN.

Contents

List of illustrations

Preface

This book had its origins, a long time ago, in my University of London PhD thesis. I thank the British Academy for funding my research, the staff of the relevant departments and libraries of the University of London, particularly the staff of the Institute of Classical Studies Library, Mark Hassall who supervised the thesis, and Lawrence Keppie and John Wilkes who examined it. Richard Alston read not only much of this book in manuscript form, but also most of the original thesis, and did much to improve both (not least, by correcting my mathematical errors). The transition from thesis to book has taken place at Cardiff University, and I must thank my colleagues there for their contributions. I owe a great debt of gratitude to Louis Rawlings, who encouraged me to think about warfare in different ways, read the entire manuscript, and made a huge difference to it. All my Ancient History colleagues at Cardiff acted as sounding boards at one time or another (sometimes at very odd times) and I am grateful for their tolerance, particularly those whose interests are far removed from warfare. Geoff Boden drew the computer graphics, Elina Brooke the illustrations, and Jon Coulston provided the photographs from Trajan's Column and Adamklissi. I am also grateful to The British Museum, Cambridge University Collection of Air Photographs, The American Academy at Rome and the German Archaeological Institute in Rome for photographs and permission to publish them. My greatest debts though are to my mother, for her support and encouragement, and to my father, in whose memory this book is dedicated, for understanding.

Kate Gilliver
Cardiff University

Introduction

The Roman army has for a long time been a subject of great interest to a wide audience and, to judge by the regular publication of new books, it is a subject that retains its popularity and fascination. The subject has its attraction partly because of the image of the Roman army as the first 'professional' army, whose organization and discipline led it to famous victories in battle and great successes in war, conquering in a remarkably short period the 'civilized' world of the Mediterranean. With the army came the famous generals who commanded it: the Scipios, Marius, Pompey and Caesar, and their skills in the art of generalship; such great commanders continue to excite popular interest today. Modern writers though are not the only ones to express an interest in and admiration for Rome's armies. There is a wide range of ancient literature on the subject too, written by both Romans and non-Romans, men who had fought for Rome and, like the Jewish writer Josephus, men who had opposed her. Because of their desire to explain the workings of the Roman army to their wider audiences, the non-Roman writers provide us with some of our best and most detailed descriptions of Roman armies and their military procedures. The works of Josephus and the earlier Greek historian Polybius express particular admiration for the discipline and training of Roman armies. They attribute much of the Roman success in war to these things, elements that we are also familiar with from modern armies.

Whilst some Roman commanders might have possessed an instinctive genius in the art of generalship and a natural ability to command and lead in war, most probably did not. They were not, however, trained in these skills in the way that their men were trained in soldiering, or as army officers are trained today. Rome had no military schools or academies and officers of the Roman army received no formal training in the arts of command, generalship and war. The senior officer corps of the Roman army was essentially 'amateur', made up of aristocratic men pursuing political careers. Aspiring politicians who were holding military commands had to acquire the necessary knowledge and skills through experience as they served as junior officers on the staff of a provincial governor or other military commander. Those who had already reached high office without the necessary skills had to rely on others who had them, or look elsewhere for the information. The Greek historian Polybius notes the importance of personal experience in the education of a general, but also the value of collections of examples illustrating the actions of earlier generals, and of textbooks.

A number of these works, contemporary textbooks on Roman military theory and manuals on field practices and pieces of equipment, survive from the Roman period. In subject matter they cover a range of material, from general works on campaigning and generalship to more detailed and specialized manuals that provide the specifications for artillery pieces, and even *How to throw the javelin from horseback*, a lost work by the prolific Elder Pliny, written whilst he was commanding an auxiliary cavalry unit in Germany. In the fourth century AD, Vegetius compiled a general treatise dealing with different aspects of training and campaigning. His is the only general treatise in Latin to have survived from antiquity. The work was a summary of earlier textbooks though, and is valuable for students of the army of the Republic and early Empire as well as Vegetius' own day, but it is not without problems. Because it is an *epitome* or abridgement of earlier works dating to the Republic and Empire, it can be very difficult or even impossible to tell to which period some of Vegetius' information is relevant. The only other general manual to have survived from the Roman period is that by the Greek philosophical writer Onasander. He wrote a textbook on the art of generalship in the mid-first century AD, paying particular attention to the personal and moral qualities of the general, and the behaviour he considered necessary for the 'good' commander. More detailed works include that by the second century AD architect Apollodorus of Damascus. His manual on siege warfare was produced to regain the favour of the emperor Hadrian, by whom he had been exiled allegedly for criticizing the emperor's architectural designs. His rough contemporary Pseudo-Hyginus, a military surveyor, wrote a manual on encamping an army on campaign. Details of these works, and the most important which have survived or are known from antiquity, are included in Appendix 1, which also lists the most easily available texts and translations of those works which are still extant.

Some textbooks, such as Frontinus' *Stratagems*, or the *Epitome of military science* by Vegetius, are well known even today, and have been influential in the past. Vegetius' *Epitome*, for example, came to be regarded in western Europe as the soldier's equivalent to the monastic rule of St Benedict, and Charlemagne expected all his officers to own a copy. Machiavelli based his ten book *Arte della Guerra* heavily on the Roman writer's work, and like Vegetius stressed the value of the citizen army and the dangers of mercenary forces, ideas that are also apparent in Machiavelli's best known work, *il Principe*. The amount and extent of influence such textbooks had on Roman military procedure, though, is much harder to evaluate. All the works claim to be educating the reader and to be providing practical advice for application in the field, but such claims are merely a part of the particular genre in which the authors were writing, and we should not accept them at face value. Claims to practical value are somewhat dubious when attached to treatises describing obsolete machinery or military systems, such as Heron's manual on the out-of-date Belopoeica artillery engine, or the detailed descriptions of the organization and drills of the Macedonian phalanx compiled under the Roman Empire by Aelian and Arrian. Many of the treatises, however, as we shall discover in the following chapters, describe and explain the arrangements and practices of Roman armies, as detailed by historians such as Polybius, Livy and Josephus, and eye-witness commentators like Caesar. Much of the guidance these treatises provide is sensible, realistic, and so practical, though at times bordering on little more than common sense.

Aspiring generals, as Polybius suggests, clearly did consult textbooks like those that have survived: Cicero is complimentary or critical of men who did so, depending on the circumstances of his comments, and the individual on whom he is passing judgement. He is prepared to praise Lucullus for studying his military textbooks in preparation for taking command of the war against Mithridates; he even jokes about a letter of advice on generalship that he had received himself from a friend when he was sent, reluctantly, to govern the Turkish province of Cilicia in 51 BC. From his praise of men like Marius and Pompey for their generalship, however, it is clear that learning through personal experience was preferred.[1]

Nonetheless, because they describe, and sometimes prescribe, the field practices of the Roman army, the textbooks can help us to understand Roman military practices and, as much as they existed, set procedures and regulations. We must remember, though, that they are textbooks and do not themselves describe the normal field practices, as opposed to an ideal propounded by theorists. Textbooks have their limitations: they cannot cover all possible eventualities, especially in something as variable as war in which luck, as the treatise writers themselves admit, plays a big part. We must then use these textbooks along with historical accounts of campaigns, and together they can help us to understand the Roman 'art of war'.

This book is intended for all interested in the Roman army and in ancient warfare, though some knowledge of Roman history and politics is assumed. All extracts from ancient authors have been translated, as have many technical terms. Latin and Greek terms that are in common usage are retained, but explanations are provided in the glossary. The themes and contents of the following chapters are dictated by the contents of the military treatises that are the principal subject of this book. These writers concentrate on the specifics of waging war and campaigning. They explain how to march an army from A to B, but not the purpose of marching an army from A to B, nor what it might do when it gets there. Pitched battle is of fundamental importance in the eyes of the treatise writers, but only as a means of defeating the enemy army at a particular moment: the more general purpose of pitched battle in the course of a war is not discussed. The theoretical writers are not interested in the causes of war, the aims of war and grand military strategies. This book, like the treatises, looks at specific aspects of campaigning, and does not claim to provide a comprehensive study of Roman warfare. It covers the composition of the army and the basics of campaigning: marching, supplying and encamping an army, the fighting of pitched battles, and siege warfare, because these are the topics of most interest to the treatise writers. The time period covered ranges from the middle Republic through to the late Roman Empire, from a period when Rome was hugely successful in war and aggressively expansionist to a time of defeat and emphasis on defence. Throughout the discussions that follow, we can see not only the close correlation between military theory and campaign practices, but also the limitations of theory and, sometimes, the acceptance by theorists of their limitations. Whilst they might be able to provide a general account of ancient military theory and certain normal procedures of Roman armies, textbooks and collections of military *stratagems* cannot of course cover everything. Experience and inspiration were also vital to the general waging successful war for Rome.

1 The army: organization

Introduction

> Military science, as the most famous Latin author puts it in the opening of his poem, consists of 'arms and men'.
>
> Vegetius 2.1

Much has been written on the organization of the Roman army and the units it comprised. Historians have made use of the considerable body of epigraphic, papyrological and archaeological evidence in the study of the army of the imperial period, along with the theoretical work of Pseudo-Hyginus. For the Republican period, however, the evidence is rather more limited. The archaeological material is restricted essentially to a few encampments in Spain, and although both Livy and Polybius provide descriptions of the Roman legion, they do not include detailed information on the allied troops fighting alongside them. None of this information, moreover, for any period, is without its difficulties. Whilst the archaeological evidence may be incomplete and difficult to interpret, the formal descriptions of the army in literary works tend to present an ideal view of the Roman army rather than the reality. Suggestions of that reality are provided by papyrological records and the recent finds of wooden writing tablets from Vindolanda near Hadrian's Wall, and hint at the sometimes considerable differences between theory and practice. Despite all the available information, however, there are still gaps in our knowledge and understanding of how the army was organized and functioned on a day-to-day basis.

This chapter will concentrate on the men of the Roman army, from the generals and officers to the soldiers and the 'non-combatants' with the army. The chapter will provide a brief description of the evolution of the two principal branches of the Roman army, the legions and the allied or auxiliary units. Armies were accompanied by camp servants and followers, the first a semi-formal part of the army, the latter consisting of 'hangers-on' providing more casual but sometimes important services to the army. Because this book deals primarily with warfare on land and not at sea, the Roman navy will not feature in this section, though the role of the navy in supply and logistics will be touched upon elsewhere.

The officers and commander

The commanders and senior officers of Roman armies had no specific qualifications for their tasks and received no training in the art of war. Throughout both the Republic and the Empire, almost all military commanders were drawn from the elite senatorial and equestrian classes, and were pursuing a political career. The military service these men fulfilled was simply one aspect of this career, along with civil offices and, in the case of the senators, magistracies in Rome. The governorship of a Roman province brought with it command of whatever Roman forces were stationed there. In the Republic in particular, this was often coupled with an opportunity or expectation to wage war against any perceived enemies of Rome; military success and glory in turn enhanced a senator's political career. In the increasingly competitive political climate of late Republican Rome, the desire of senators to obtain important military commands and the accompanying prospects of wealth and glory from successful campaigns helped to encourage an aggressive foreign policy. Even Cicero, whose military experience was limited to brief appointments as a junior officer on the headquarters staff in the early first century BC, was eager to campaign and obtain a military triumph when governor of Cilicia in 51 BC. A relatively minor campaign against some Cilician tribes brought success and the desired acclamation as *imperator* from his soldiers.[2]

In the Republic, there were no officially appointed commanders of legions or cavalry units: the magistrate or provincial governor commanded the army in his province and it was up to him to appoint subordinates as required. Those appointed were likely to be on the staff of the general, drawn from his friends, relatives, colleagues and clients. Cicero had his brother Quintus with him during his governorship of Cilicia, and left him in command of the whole army at the end of his brief campaign. Marius and Rutilius Rufus, who had both held the quite senior magistracy of praetor in Rome, served as subordinates to their patron Metellus in the war against Jugurtha, and both at times successfully commanded parts of Metellus' army. Such appointments were not always so fortunate though: one of the early commanders in the Jugurthine War, Spurius Albinus, handed command to his brother Aulus over the winter. Aulus decided to continue the campaign despite the season, led his army into an impossible situation and was forced to surrender to Jugurtha to preserve the army. In such appointments social status was of considerable importance. Plutarch reports how during the civil war between Caesar and Pompey, Domitius Ahenobarbus appointed a man of no military experience to a responsible post because he had an agreeable personality and sound character.[3]

In the Empire, commanders of legions and auxiliary units were normally appointed by the emperor, as were the governors of all the provinces that contained significant military forces. Some provincial governors during both the late Republic and the Empire might, like Cicero, have had very little or no military experience before their appointment. Some historians have dismissed concerns about the lack of practical experience of these commanders and officers, pointing instead to the legionary centurions who are often seen as a core of 'professionals' from whom inexperienced equestrian and senatorial commanders might learn.[4] Whilst many centurions clearly had considerable experience of military affairs, this was not always the case. In the Imperial period, men of sufficient social status might be appointed directly into the centurionate and therefore might have had no more relevant military experience than some equestrians and senators. Most governors of important military provinces, however, had previously acquired military experience. A common pattern in the early Empire was for a senator to perform service as a military tribune for some months in his early twenties, and then for several years as legate of a legion in his mid to late thirties. Pliny the Younger may have spent his military tribunate auditing the account books of auxiliary units in Syria, but according to Tacitus, Agricola had spent his learning the rudiments of campaigning and generalship.[5]

When listing the qualities and qualifications of a good general, the Roman writers include good character, reputation, integrity and oratorical skills, all of which point to a man of high social status. Ability at warfare was only one of the principal attributes that Cicero felt a perfect general should possess, the others being prestige, luck, and knowledge of military affairs. The latter involves an understanding of military theory, which could be obtained from the textbooks that writers such as Polybius believed could teach men the art of generalship and campaigning. Though practical experience was preferable and more worthy of praise, it was accepted that such textbooks did have a role in the education of generals.[6]

The army

For many ancient writers, the legion formed the backbone of the Roman army. Vegetius saw the legion alone as being the principal reason for the acquisition of Rome's vast empire. He bemoans the demise of the traditional legion or *antiqua legio*, criticizing the legion of his own day for lacking discipline and being ineffective, and the auxiliaries for being unreliable. Roman historians tended to concentrate more on the actions of legions in campaigns and pitched battles than the Italian allies (the *socii*) or auxiliaries (the *auxilia*). Tacitus points out that it was very hard to keep track of auxiliary units because of the sheer number of them. Even so, there is more emphasis on the actions of the citizen soldiers of the legions than the non-citizens that made up the *socii* and the bulk of the auxiliaries. This emphasis is explained partly by Tacitus' comment, but is also because the legion was composed of Roman citizens, and in the Republic in particular, the citizen body comprised both army and state. The *socii* or allies were literally an *auxilia*, additional troops to supplement the legions, and in the eyes of ancient historians, the citizens and legionaries were superior. The idea of the dominance of the legion continued in Latin literature even though the auxiliaries were as important and valuable in warfare as the legionaries.[7]

Polybius and Livy provide detailed descriptions of the organization and armament of the Republican legion, and its methods of deploying and fighting. Although they both concentrate on the legion, they also include some information on the nature and role of the allied forces or *socii* in the Republican period. Unfortunately, there are no equivalent descriptions for the imperial period, and the sources are less reliable. Josephus' brief excursus on the Roman army includes details of unit size, but his numbers are problematic and idealised. The most detailed description of the army of the imperial period is that provided by Vegetius, but this deals exclusively with the legion. As with much of Vegetius' work, moreover, the information he provides is by no means straightforward. Vegetius does not indicate which sources he was using for this section of his *Epitome*, and it is not clear to which period, if any, his description of the legion belongs. The description is a necessary precursor to Vegetius' third book, which deals with the deployment and role of the legion in the field of battle, but it is confused and contains mathematical errors. The work of Pseudo-Hyginus is clearer and, for the early Empire, probably much more accurate.

Pseudo-Hyginus' treatise on encamping an army includes, of necessity, information on the size and internal organization of both legions and auxiliary units, but this is brief. Since the author was writing a handbook for other military surveyors, who would presumably have been aware of the usual organization of army units, a detailed description is unnecessary. He therefore confines himself to providing only the information directly relevant to his work. He is not interested in how each unit was armed and functioned in battle, but only in the amount of camping space each type of unit or category of soldier required. The first part of the text, which would have given the details of legionary organization, is missing, but the brevity of the author's descriptions of the auxiliary units later in the work suggests that it would not have been lengthy. Other historians, such as Sallust and Caesar, were writing for an informed audience and included very little detail on the intricacies of unit organization. They are much more useful in explaining how the different units actually functioned whilst on campaign and, to a certain extent, in battle.

The legion

Maniples

The manipular legion probably came into existence during the early to mid fourth century BC to replace the phalanx previously used by the Romans. Although the creation of the legion is traditionally attributed to Camillus in the aftermath of the capture of Rome by the Gauls in 390 BC, there is no strong evidence to indicate when and by whom the legion was actually introduced. It is more likely that a gradual development occurred from phalanx to legion rather than a sudden reform, but the manipular legion was probably functional by 340 BC, the point in his history at which Livy describes its organization and workings. Livy includes this description to help explain the nature of the war between the Romans and their former Latin allies who had revolted, and he implies that the army had been organized in this fashion for some time. Many historians, including Keppie, suggest that the manipular legion was developed to deal with the looser fighting formations used by the Gauls to such effect against the Romans in the late fifth and early fourth centuries BC. But a number of ancient writers suggest that the Romans adopted the rectangular shield or *scutum* and the manipular formation from the Italian tribe of Samnites. It is well known that the Romans did adopt some of the equipment and fighting techniques used by their enemies, such as the Spanish sword, and adapt them for their own use. This is something that many Greek and Roman writers noted, and it became something of a literary tradition to suggest the foreign origins of different Roman weapons and military practices. The suggestion that the *scutum* and manipular organization were of Samnite origin is part of this literary tradition and we should be wary of accepting it as reliable.

Even though we remain uncertain of its origins, however, the literary traditions and what we know of the development of the legion indicate that the Romans adapted their fighting styles and formations to deal with different types of enemy.[8]

Polybius' famous excursus on the Roman military system provides a clear account of the organization of the manipular legion. Although he was writing in the mid second century BC, his account, possibly taken from an officer's handbook, probably dates to an earlier period, perhaps from the time of the Second Punic War. Polybius' description of the Roman legion's organization is quite clear, but it is Livy who explains how the legion functioned in battle, at least in theory.

The organization of the legion according to Polybius (**1**):

Soldiers	**Equipment**
1200 *velites* (light infantry)	sword, javelins, small shield, helmet
1200 *hastati* (heavy infantry)	Spanish sword, two *pila*, *scutum*, helmet, greaves, breastplate or mail coat
1200 *principes* (heavy infantry)	armed as *hastati*
600 *triarii* (heavy infantry)	armed as *hastati*, but with spears (*hastae*) instead of *pila*.
300 cavalry	

The *hastati* and *principes* were each divided into 10 maniples of 120 men, the *triarii* into 10 maniples of 60 men; each maniple was assigned two centurions, the senior of whom commanded the maniple. The *velites* were divided equally amongst all the maniples, and the cavalry were divided into 10 *turmae* commanded by decurions. The legionary strength according to Polybius' information is 4200 infantry and 300 cavalry, but this was not necessarily fixed, and the legion might be increased in size in times of emergency, or if a stronger force were needed. If the legion was larger, the extra troops would be divided among the *velites*, *hastati* and *principes*; the number of *triarii* remained constant. The cavalry force might also be increased.

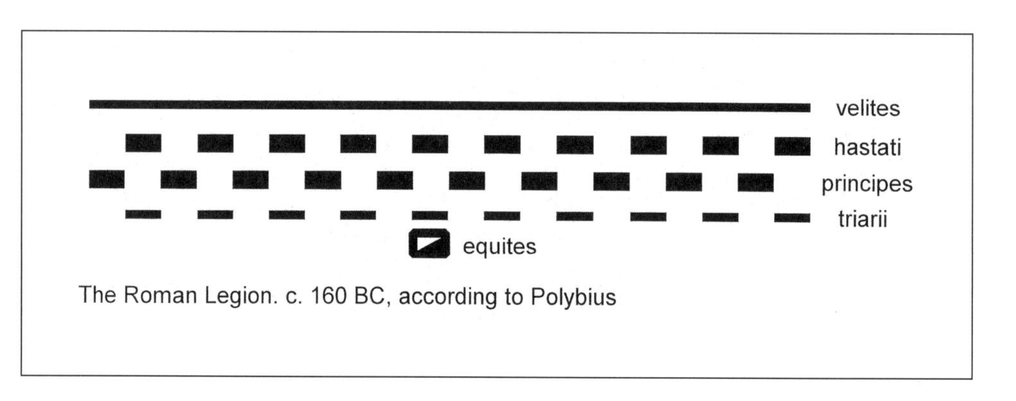

1 The organization of the 'Polybian' legion

Livy's description of the legion is similar, but here the rear line is much more complicated than Polybius', which contained just maniples of *triarii*. Livy's line comprises not just *triarii*, but also *rorarii*, whom he calls 'younger and less experienced men' and *accensi*, described as the men of least reliability. This conflict with Polybius' description is confusing, particularly since other sources indicate that *rorarii* were light-armed troops who, like the *velites*, commenced the battle, and *accensi* were camp servants.[9] Livy mentions only the *triarii* in the rear line when explaining the fighting technique of the legion; there is no trace of either *rorarii* or *accensi*. The most likely explanation for this inconsistency between descriptions is that Livy used two or more different sources, which described the legion at different stages of its development, but conflated them to produce one confused account. Livy records legionary strengths of between 4000 and 5000 during the Hannibalic War, and it is clear that, as indicated above, legionary strength could fluctuate, depending on the military situation. The legions at Cannae were nominally 5000 strong with 300 cavalry, in 216 BC legions of 5000 with 400 cavalry are reported, and Livy claims that the legions Scipio took to Africa were 6200 strong, with 300 cavalry, though this may be an exaggeration. By the second century BC, the normal size of the legion may have been increased to c.5000, with 6000 in times of military necessity, such as those raised for the war against Perseus in the late 170s BC. Figures from the later Republic though indicate that legions could be considerably under strength. At Pharsalus, Caesar's cohorts had an average strength of only 275 (instead of the theoretical strength of 480), and soon after the battle his 6th Legion was down to under 1000. Long service, sickness and battle casualties had reduced his legions' strength, and Caesar was unable to obtain sufficient recruits to bring his units up to their theoretical strength.[10]

Livy explains how the 'manipular' legion functioned:

> When an army had been drawn up in this order, the *hastati* were the first to enter battle. If the *hastati* failed to overcome the enemy, they retreated slowly and were received through the gaps in the line of *principes*. Then the *principes* took up the battle, with the *hastati* behind them. The *triarii* knelt beneath their standards with their left legs stretched forward, shields leaning against their shoulders and their spears thrust into the ground and pointing forwards, so that the battle line bristled like a protective palisade. If the *principes* too were unsuccessful in their fight they gradually retreated from the front line to the *triarii* (from this comes the saying 'to have reached the *triarii*' when things are going badly). Once they had received the *principes* and *hastati* through their ranks, closing the lanes, so to speak, with no more reserves behind to rely on, the *triarii* fell on the enemy in one continuous line. This was a particularly frightening thing for the enemy who were pursuing men whom they believed they had defeated, only to see a new battle line suddenly rising up with increased numbers.
>
> Livy 8.8

He provides no specific details of how the different ranks carried out the manoeuvres, but there seems no reason to dispute his basic description. The manipular legion contained troops armed and equipped differently, who carried out different functions in pitched battle. The 300 strong cavalry contingent attached to each legion added versatility, though it was Rome's allies who usually provided cavalry in strength.

Once established, the legion did not have a static organization, formation and method of fighting. As with the introduction and development of the legion, the necessity to campaign in different terrain and face new adversaries with different armaments and ways of fighting led to changes in the legion. Economic and political factors also influenced the way the army was organized and equipped. Because the surviving sources provide so little detail on such matters, though, it is very difficult to know when and how changes took place. Even some of the more radical changes, such as the organizational change from maniples to cohorts in the late Republic, are very poorly documented, but changes seem to have been gradual developments rather than sudden reforms.

Cohorts

The writer Cincius Alimentus produced a treatise in the early Empire in which he stated that the legion was made up of 60 centuries, 30 maniples and 10 cohorts. This is one of the earliest references to the legionary organization that historians have traditionally attributed to the military reforms of Marius in the late second century BC, by which the cohort rather than the maniple became the principal tactical unit of the army. The transition from manipular legion to that based on the cohort took place at some time between the Second Punic War, for which we have good evidence of the earlier organization from Livy and Polybius, and the mid first century BC when Caesar was campaigning in Gaul. Caesar's commentaries on these campaigns illustrate clearly the use of the 'cohortal' legion rather than the manipular. Marius did make several changes to the armies under his command, among them the introduction of the legionary eagle to encourage *ésprit de corps*, but there are no strong grounds for crediting him with the reform of legionary organization.[11]

Livy and Polybius report cohorts of legionaries operating in Spain during the Second Punic War. There seems no reason to doubt these reports despite the fact that the legion was at this time usually organized in maniples and fought pitched battles using the manipular formations described above. These legionary cohorts were probably introduced specifically for the fighting in Spain. The nature of the mountainous terrain and the fighting, against Spanish tribes using dense infantry formations and effective cavalry, and sometimes operating a form of guerrilla warfare, required units smaller and more flexible than the legion, but larger than the maniple.[12] There is no indication, however, that from this time legions in Spain were always based around cohorts, or that cohort-based legions were used in other provinces before the universal adoption of this form of legionary organization. Sallust includes some of the latest references to manipular legions in his narrative of the Jugurthine War in Africa, fought from 112-105 BC, but like many Roman historians he is careless in his use of military terminology. Thus, we find him referring to light-armed cohorts of legionaries as well as, later in the work, light-armed maniples of legionaries. He also uses these terms when referring to Jugurtha's Numidian forces, which as far as is known were not organized along the Roman model. Sallust does not bother to differentiate between different types of organization: he is simply using the words *cohort* and *maniple* to indicate a unit of men, probably smaller than a legion. Tacitus

2 Legionaries of the first century AD, Mainz. Courtesy Landesmuseum, Mainz

does the same. Writing of Roman legions that were without any doubt organized in cohorts, he nonetheless regularly uses the term *maniple*; for him it seems to be virtually interchangeable with the word *century*, denoting a unit of small but indeterminate or unimportant size.[13]

Despite the problems of dealing with the inaccuracies of Sallust's vocabulary, it is very likely that the final change from manipular to cohortal legions did occur in the late second to early first centuries BC, though it is not until Caesar's Gallic campaigns of the 50s BC that the literary sources provide details of the cohortal legion on campaign. By this time too, the legionaries were armed in a uniform manner, with mail coat and helmet, *scutum*, *pila* and Spanish sword, all well attested from this period onwards in the archaeological evidence. The *triarii* and the *velites* have vanished. The move towards uniform arming and equipping of the legion was probably influenced by military considerations, but may also

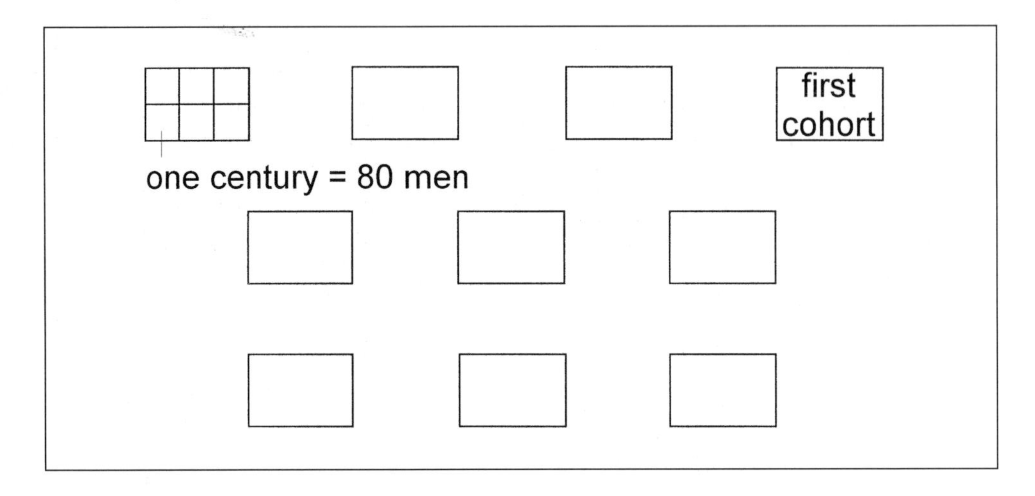

3 *The imperial legion: ten cohorts in a 4-3-3 formation*

have been encouraged by economic factors. The reforms of Gaius Gracchus in 123-2 BC provided legionaries with their arms at the state's expense whereas before this the men equipped themselves. This measure would have increased in importance after Marius started recruiting in numbers from the *proletarii*, the Roman citizens who possessed no property and so could not have provided their own arms. It may simply have been easier and cheaper to provide all the soldiers with the same equipment.

There is a considerable body of evidence, literary, epigraphic and archaeological, about the size and organization of the 'cohortal' legion. Although the principal features of this legion are well attested, there are still some details that are little known or understood. The legion contained ten cohorts, each of which was subdivided into six centuries of 80 men. A small cavalry unit of about 120 was drawn from the infantry, probably for scouting and carrying messages. Since they remained 'on the books' of their original centuries, they may well have been included among the 80 men in each century. The nominal strength of this legion then was 4800 **(3)**. In his brief comments on legionary strength, Pseudo-Hyginus states that the first cohort of the legion was *milliary*. This term literally means one thousand strong, but like the century which contained only 80 men and not the hundred the name implies, the *milliary* cohort probably contained about 800 soldiers in five double centuries rather than 1000, giving the legion additional strength at about 5120. Vegetius also mentions this milliary first cohort, and it is attested by archaeological evidence.[14]

The legionary fortress at Inchtuthil in Scotland appears to illustrate perfectly the legionary organization that Pseudo-Hyginus describes **(4)**. Excavated in the 1950s and 60s, the plan of the fortress shows nine groups of barrack blocks, each of which could house the six centuries of an ordinary or *quingenary* legionary cohort, and a further group of barracks next to the legion's headquarters building (the *principia*) in the centre of the fortress. This group of ten barrack blocks and five houses has been identified as the accommodation for a milliary first cohort with its five centurions. A number of other fortresses, including Caerleon and Chester, have sufficient space to one side of the

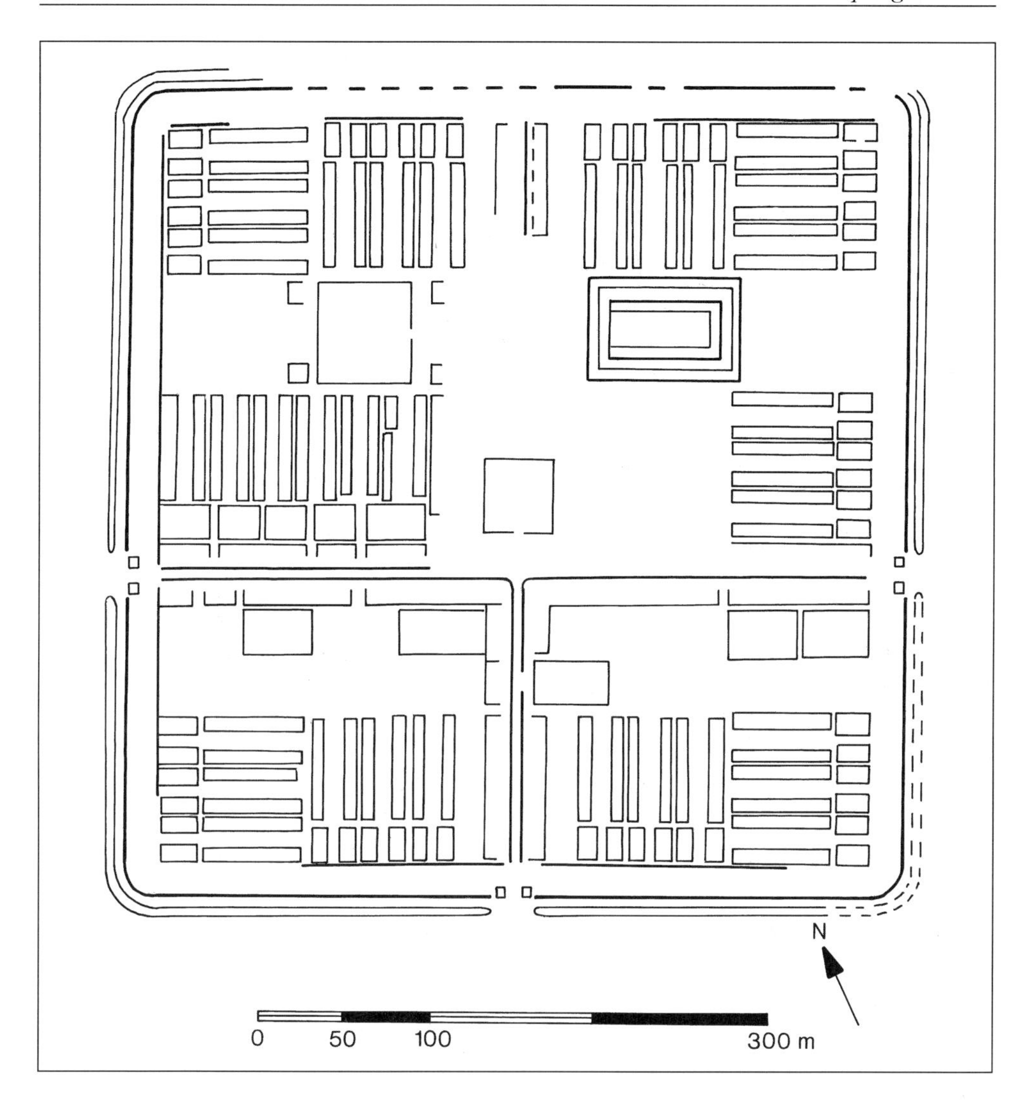

4 *The legionary fortress at Inchtuthil (after Pitts & St Joseph)*

principia to accommodate a milliary first cohort, but the excavations have revealed only barrack accommodation for a first cohort of the same size as all the other cohorts, with only six normal-sized centuries. These stone barrack blocks date to the second century AD, by which time the first cohort may have been reduced in size.

It looks as if at some point the first cohort was made milliary, but was then reduced to normal or quingenary size, and it is this reduced size first cohort that is reflected in the remains of barrack blocks at most legionary fortresses. This 'reform' of enlarging the first cohort is usually dated to the Flavian period to tie in with the evidence from Inchtuthil and elsewhere. However, there is no reason to suppose that all legions throughout the empire had milliary first cohorts during this period. Legionary first cohorts may simply have been enlarged in provinces where legions were likely to have been heavily involved in campaigning, such as Britain in the Flavian period (the period in which Inchtuthil was constructed). The strength of the manipular legion could be increased in times of military necessity, and this is likely to have been the case for the cohortal legion too. It was far easier and quicker to increase the size of units for campaigns or emergencies than to raise new ones. Certainly auxiliary units in the imperial period varied in size according to military need, and it is probable that the legion continued to do so as well.

The normal deployment of the cohortal legion was as either a *triplex acies* or triple battle line in 4-3-3 formation, or a *duplex acies*, a double battle line with two lines of five cohorts. But the system allowed considerable flexibility and cohorts could easily be detached for service elsewhere in the battle line, or as reserves, or in ambush (see Chapter 4 on pitched battles). Despite the tactical importance of the legionary cohort, there is some dispute about who commanded it. It may have been the senior centurion in the cohort, in the same way that the senior centurion in the maniple commanded that unit. However, Vegetius states that the cohorts were commanded by tribunes or other officers nominated by the *princeps*, probably meaning the Emperor. These appointments may have been on a fairly informal basis, in much the same way as commanders of individual legions in the late Republic.[15]

Allies and *auxilia*

A Roman army on campaign always included a complement of allies. The provision of allied troops for military service was one of the most important features of Rome's treaties with conquered tribes in Italy and elsewhere, and the allied contingent was usually the same size as the legionary, if not larger. Allied cavalry contingents were considerably larger than the Roman, and Rome relied very heavily on others for her cavalry forces. In the Republic, additional forces or *auxilia* might be recruited locally to supplement forces further, particularly if the Roman army was campaigning overseas. Such troops might have local knowledge of topography and the enemy, and could provide specialist fighting techniques appropriate to the situation. Not a great deal is known about the size or organization of contingents of allies (*socii*) or *auxilia* in the Republican period. Livy assumes that the Italian allied infantry was armed and fought in the same manner as the legionaries, and this may well have been the case with some allies, but not necessarily all. Polybius provides very little detail on the allies in his excursus on the Roman army, but does say that they formed the two wings of the army, the *ala sociorum*. It seems likely that the individual allied contingents were led by their own leaders, drawn from their local aristocracies, but the allies as a whole, both infantry and cavalry, were commanded by Roman officers.

As well as expertise in cavalry, the units of *auxilia* provided troops of a specialist nature, such as archers and slingers and most famously in the imperial period, the Batavians who were renowned for their water-borne assaults. The units of *auxilia* increased in importance during the late Republic for several reasons. Firstly, the phasing out of the *velites*, the light armed troops in the legion, and the uniform arming of the legion, left Rome having to look elsewhere for light infantry for tasks such as skirmishing and pursuit of the enemy.

5 *Legionaries with their equipment, Adamklissi.* Courtesy JCN Coulston

After the Social War of 91-87 BC, Rome's Italian allies, the *socii*, were granted Roman citizenship and therefore became eligible to serve in the legions, abolishing both the infantry and the cavalry that the *socii* had been providing. By this period, Rome was already making much more use of cavalry recruited from outside Italy, such as Gaul and Numidia. The Numidian prince Jugurtha, for example, commanded a unit of Numidian cavalry in the Roman army at Numantia in 133 BC. However, the political and military changes during the late second to early first century BC meant that all non-legionary troops now

6 *Auxiliary of the first century AD, with shield and spears, Mainz.*
Courtesy Landesmuseum Mainz

had to be recruited from *auxilia* as part of tribute, or through friendships and alliances. Such troops were usually raised by an individual Roman general for the duration of a campaign, and the literary sources sometimes raise questions about the expertise and commitment of some of these troops, though others are portrayed as highly effective. It was not until the early Empire that the auxiliary units began to be retained on a permanent basis, with set terms and conditions of service. It is from the early Empire too that better evidence survives about the size and organization of these auxiliary units.[16]

Pseudo-Hyginus provides the most detailed information on the size and organization of auxiliary units. There were three different types of unit, infantry cohorts, cavalry units and 'part-mounted' cohorts which contained both cavalry and infantry; all types of unit could be either *quingenary* (nominally 500 strong), or *milliary* (nominally 1000 strong). Pseudo-Hyginus' figures may be tabulated thus:

	Quingenary	men	Milliary	men
Infantry cohort	6 centuries of infantry	480	10 centuries of infantry	800
Part-mounted cohort	6 centuries of infantry	480	10 centuries of infantry	800
cohors equitata	120 cavalry	120	240 cavalry	240
	Total	600	Total	1040
Cavalry unit	16 *turmae* of cavalry[17]	480 or	24 *turmae* of cavalry	720 or
Ala		512		768

It is important to remember that as a theoretician Pseudo-Hyginus gives the 'paper' strength of the units rather than their actual strength. Documentary evidence in the form of papyri and wooden writing tablets has shown that in practice the strength and organization of auxiliary units varied considerably. Cohors I Tungrorum, stationed at Vindolanda in the early second century AD, was nominally a quingenary infantry cohort but was considerably over strength with 761 men on its roster, though fewer than half were actually available for duties at the fort on the date of the roster. On the other hand, a quingenary part-mounted cohort stationed in Egypt in AD 156 was under 'paper' strength with only 363 infantry, 114 cavalry (and 19 camel riders). Rosters for other units show similar irregularities, with cohorts under or over their theoretical strength. It is not surprising to find that units, either legionary or auxiliary, had an actual strength that differed from the theoretical strength. We have seen how legions could be increased in size during the Republic in times of crisis. Under such circumstances, the demands on the *socii* might also be increased, and there is every reason to suppose that the same could happen during the imperial period with both auxiliary units and legions. It is quite likely, indeed, that units were kept understrength much of the time, and could be brought up to strength as the political or military situation required.[18]

Although their strength varied, these auxiliary units were retained in existence on a permanent basis, unlike the units of *socii* in the Republic. In addition, Roman generals continued to recruit local troops on an *ad hoc* basis when campaigning in certain parts of the Empire. In minor campaigns in Thrace under Tiberius, the Roman commander Sabinus was joined by troops raised by the local king Rhoemetacles for the duration of the campaign, and Corbulo used local levies of Iberians in his campaign in Armenia. He used their local knowledge of the terrain and enemy, sending them against the neighbouring Mardi in mountainous country while he deployed his regular Roman troops elsewhere. Arrian too had local levies in his campaign against the Alans, heavy infantry from Lesser Armenia and Trapezus, and spearmen from Colchis and Rhizus on the Black Sea.[19] These troops provided additional assistance to the Romans, but they were not used to bear the brunt of pitched battle in the way that legions or regular auxiliary units did.

Some modern writers have suggested that particular auxiliary units may have been attached to a specific legion as part of a 'legionary command'. Tacitus may hint at this when he describes eight cohorts of Batavians as the auxiliaries of the Fourteenth Legion, and the legionary fortress at Neuss in Germany probably held auxiliary troops in addition to the legion stationed there in both the Claudian and Flavian periods. Much of the second stone phase of the fortress has been excavated: the original stone fortress of the Claudian period had been destroyed during the Batavian revolt and was rebuilt in the early 70s by Legion VI Victrix. The plans of the Flavian fortress show additional barrack blocks above the number required for the legion, and these probably housed an auxiliary unit, possibly including some cavalry. Other fort plans suggest a joint occupation of auxiliary units and legions or detachments of legions, such as Newstead in Scotland during the Antonine period **(7)**, but there is nothing to indicate that particular auxiliary units were permanently attached to legions. Their shared accommodation may simply have been a matter of convenience, or the strategic importance of the location, rather than being indicative of any administrative relationship between the units.[20]

Size and composition of armies

Vegetius extols the virtues of an army of fairly compact size. On several occasions he claims that a small well trained and well disciplined force will defeat a poorly trained and badly organized army despite being vastly outnumbered. He notes that 'the Roman people conquered the whole world with its military drill, camp discipline, and military skill'. Other military writers such as Onasander and Frontinus stress the importance of drill and discipline, as does Polybius in his excursus on the Roman army. Modern commentators too have placed considerable emphasis on the organization, training and discipline of Roman armies rather than their size, when considering the reasons behind their success.[21]

Vegetius compares the relatively small Roman armies favourably with the enormous armies of kings like Xerxes, Darius and Mithridates. Over-large armies, he points out, are more likely to suffer mishaps. Food and water supplies become much larger problems, marching columns are dangerously long and particularly vulnerable at river crossings; a very large army that suffers defeat will face higher casualties and be more demoralized than a small one. Vegetius stresses the enormous labour involved in collecting fodder for large numbers of pack animals and horses. Larger armies were more likely to suffer from difficulties in their supply lines and, he indicates, sufficient water might also be lacking. He might have added that smaller armies were more likely to be able to support themselves partially through requisitions from local communities or 'living off the land'. The larger the army, the smaller the proportion of its supplies it would have been able to obtain locally. Drawing on a source of Republican date, probably Cato's *de re militari*, Vegetius indicates the size of 'standard' armies of that period. A praetor on a small-scale campaign would, he says, be given a legion with allied infantry and cavalry totalling 10,000 infantry and 2000 cavalry. His estimate of the consular army roughly agrees with that of Polybius, two legions, allied infantry and cavalry to a total of 20,000 infantry and 4000 cavalry. Like Polybius, Vegetius indicates that the two consuls might campaign together

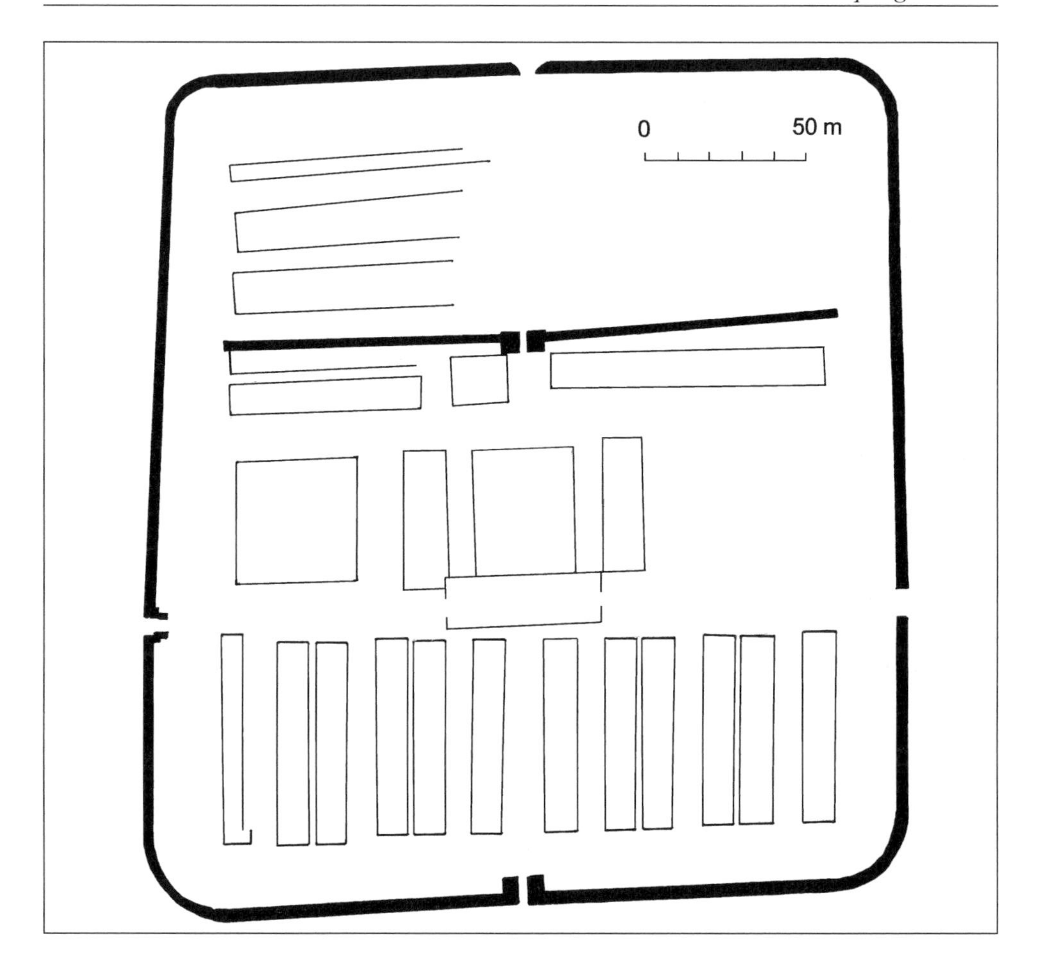

7 *The Antonine fort at Newstead. The stone wall built across the middle of the fort probably served to divide the two different units stationed there*

with an army based around four legions.[22] He suggests that the legionary contingent should not be outnumbered by the allied, seemingly for the same reason that Pseudo-Hyginus places the legions closest to the defences of his marching camp:

> Because they are the most trustworthy of the provincial units, the legions should camp next to the rampart to guard it and by their number to hold inside a human wall the army raised from foreign tribes.
>
> Pseudo-Hyginus 2

There is not a great deal of evidence to suggest that for the most part the Romans were constantly concerned about the loyalty of their allies and auxiliary units. Revolts and defections to the enemy did occur but were comparatively rare. Indeed, the deployment of allied and auxiliary troops in key tactical points of battle lines and the high reliance on the cavalry they provided indicates the trust Rome was willing to place in these troops.

The proportions of allied or auxiliary soldiers to Roman troops though did traditionally concern Roman writers. Studies of military manpower in the late Republic have indicated that *socii* in the Roman army at times outnumbered the citizen legionaries. Although the proportions of *socii* to citizens fluctuated between the Second Punic War and the Social War (218-89 BC), historians such as Velleius Paterculus are probably right when they claim that by the end of this period there were two *socii* for every citizen soldier. Vegetius' assertion that the Roman contingent was never outnumbered illustrates the tendency we noted earlier of the literary sources to emphasize the role of the citizen soldiers above the non-citizens. It is also an ideal, reflecting the views of his Republican source as well as the concerns of Vegetius' own day about the numbers of 'barbarian' or Germanic troops in the Roman army after the defeat at Adrianople in AD 378. The histories of the early Imperial period show little concern for such issues though, but continue the tradition of emphasizing the role of the citizen legions in battle.[23]

Numbers in ancient sources relating to both army sizes and subsequent casualty figures are highly problematic. The size of armies and casualty figures can be inflated by historians for dramatic effect, or reports altered for political reasons, and there can be a wide degree of variety in the numbers given in different accounts of the same event. The Roman losses at Arausio in 105 BC, where they were defeated by the Cimbri and Teutones, range from 60,000 to 120,000. It is apparent though, that small Roman armies, greatly outnumbered by the enemy, regularly brought off impressive victories with very low casualty figures. Sulla, it is claimed, defeated Mithridates' army of 110,000 at Chaeronea with an army of 16,000, whilst Dio suggests that Suetonius Paulinus' small army (Tacitus puts it at 10,000) defeated the 230,000 Britons with Boudica. Whilst these numbers are clearly exaggerated, it is equally apparent that Vegetius and the other military writers are right when they point out that Rome's usually small armies were frequently successful against much larger forces.[24]

The accounts of campaigns and battles in the imperial period rarely offer sufficient detail of the numbers and types of troops involved for detailed study of the composition and balance of the forces, though two do allow some attempt at a fuller reconstruction of the composition of the armies. For his campaign against the Alans with their very mobile force and strength in cavalry, Arrian had a balanced army of some 10,000 infantry and 3500 cavalry. The infantry contingent included approximately 7000 legionaries and some local levies of heavy infantry and spearmen. The army as a whole contained a significant force of archers, both foot and mounted. Together with the artillery he proposed to deploy on high ground and the *pila* of the legionaries, Arrian was able to face the Alans with formidable fire-power. His army was therefore suited to the nature of the campaign and the enemy; his tactics will be considered in more detail later. Germanicus' army for his campaigns in Germany in the early first century AD, included a very large force of eight legions supplemented by two cohorts of the Praetorian Guard, present because the commander was a member of the imperial family. Germanicus' auxiliary forces included cohorts of Gauls and Germans, local levies, cavalry and, like Arrian, foot and mounted archers, but Tacitus gives no details of numbers. Although Germanicus had a much higher proportion of heavy infantry than Arrian seems to have had, he was facing primarily infantry rather than cavalry, and so needed strength in this area. His auxiliaries provided him with considerable versatility, which he used successfully in pitched battle at Idistaviso in AD 16.[25]

Servants and camp-followers

A Roman army on campaign did not consist only of soldiers and their baggage and equipment. Servants attached to the army had an established role that was normally concerned with the baggage train, and supplies of food and fodder, but could involve some military action, though they were not normally armed. Merchants and other army followers such as soothsayers and prostitutes did not have the formal role that the servants had but nonetheless provided a range of services to the army and soldiers. The numbers of these servants and followers could swell the size of the army quite considerably, possibly contributing to slow progress of the army and supply problems. They were viewed by some generals as having a detrimental effect on the morale and discipline of the army. Scipio Aemilianus ejected from the Roman camp at Numantia the merchants, prostitutes, soothsayers and diviners whom he considered were demoralizing the defeated Roman soldiers. When he took over the war against Jugurtha, Metellus went further and sent away from the camp not only the merchants who had been selling the soldiers food but also the servants whom his soldiers retained. Marius made his soldiers carry their own kit instead of making their servants carry it, but as we shall see, he was willing to make use of the servants when necessary.[26]

It is impossible to make any estimate of the number of servants, merchants and other camp-followers who might have accompanied Roman armies on campaign. Tacitus claims that the armies of Vitellius and the Flavians in the wars of AD 69 contained as many camp servants and followers as soldiers. He probably exaggerated these numbers, but the instability in the empire, particularly Northern Italy during this period may have encouraged many civilians to attach themselves to the army in the hope of profiting from the campaigns. It is likely that the numbers of camp servants were usually considerably smaller. Each legion may have had some 400-600 servants, but the evidence is extremely poor and we can only guess at the numbers.[27]

Tacitus presents a rather dim view of camp servants and those who followed armies. He claims that those in Vitellius' army were more ill-behaved than slaves, whilst those of the Flavians in the second sack of Cremona are described as 'even more perverted in their lust and savagery' than the soldiers. He blames camp servants for mutilating and impaling the head of Galba after his assassination in AD 69. Tacitus however is keen to stress the breakdown of discipline and morality generally because of the civil war and anarchy of this period. He presents the rank and file soldiers in a very poor light as well and given the low social status of the servants, it is perhaps not surprising that we are given such a negative image. Other historians present the camp servants in a more positive light and explain more about their role in the army. In his description of the order of march, Vegetius states that the baggage train and its grooms were marshalled under standards and under the charge of experienced camp servants. Some level of organization and experience is suggested in Caesar's description of an attack on foragers from Q. Cicero's camp in Gaul in 53 BC. The foragers, infantry and camp servants along with cavalry were attacked by German cavalry a short distance from the camp and there was considerable confusion amongst the Roman troops. The servants showed a basic military sense by making for the nearest high ground and then, when dislodged, joined the legionaries who were forming up. Then they followed the veteran soldiers in a charge through the enemy to the safety of the camp.[28]

On other occasions, the camp servants were given a role to play in battles, even though they were not usually armed. There are several recorded episodes of servants being ordered to join in with the battle cry so as not to betray to the enemy the small size of the Roman force. Mounted on mules, they could also be used to deceive the enemy into thinking reinforcements had arrived.[29] In this role they might even join in the ensuing battle, as ordered by Marius against the Germans at Aquae Sextiae in 102 BC:

> Marius sent Marcellus at night with a small force of cavalry and infantry to the rear of the enemy, and to give the impression of a large force he ordered the grooms and camp servants to go armed, and a large number of mules wearing saddle-cloths, so that they would appear to be cavalry. He ordered them to attack the rear of the enemy as soon as they noticed battle had been joined. This plan struck such terror into the enemy that the frightful Germans turned in flight.
>
> Frontinus, *Stratagems* 2.4.6

In an attack on a Carthaginian encampment in Spain in 209 BC, the servants assisted the Romans in hurling stones at the defenders and, despite coming under heavy fire themselves, helped the Roman soldiers to achieve their objective. The bravery of servants is illustrated in Caesar's account of the battle against the Nervii. The fighting had been close and the servants, who had gone out to plunder when they thought the Romans were getting the upper hand, then fled when they saw the Nervii in the Roman camp. Later though, when reinforcements arrived and the Nervii started panicking, the servants turned on the Gauls despite not being properly armed.[30]

Camp servants do not seem to have been armed under normal circumstances and their principal roles did not normally include fighting. As we have seen, they became involved in plundering after battles, but the literary accounts do not indicate that they received any formal share in the booty. Care and control of the baggage animals seem to have been one of their major functions, along with seeing to their feed and pasturing. They may also have acted as servants for the soldiers, but soldiers may also have been accompanied by slaves. Slaves are sometimes mentioned and depicted on tombstones of the imperial period but there is little other evidence for them. Although they were a semi-official part of the army, the servants could be dismissed at any time. The primary reason for servants of free status travelling with the army is likely to have been the prospect of plunder in the event of battle, and looting of civilians.

Merchants with the army could be a useful source of additional supplies, particularly if the army was camped for some time, as at Numantia on the arrival of Scipio and in Numidia when Metellus arrived. But they were not an official part of the army and their principal reasons for accompanying an army were no doubt opportunities for profit and the establishment of new trade. Roman merchants were established in Gallic settlements very soon after Caesar's conquest, such as those in Cenabum who were massacred during the Gallic revolt in 52 BC. Whilst the servants attached to the army seem to have camped within the fortifications of the camp, the merchants and presumably any others following the army, do not seem to have done. They followed the army at their own risk, sometimes a considerable one. When Quintus Cicero's camp was attacked by the German cavalry, the merchants who were camping below the ramparts were unable to get away in time. The servants, camp followers and families accompanying the army of Varus in AD 9 were slaughtered along with the soldiers.[31]

Conclusions

> The Romans must be admired because although they dearly love their own institutions, they are willing to select useful things all around and adapt them to their own use.
> Thus you may find that they have taken certain weapons from other peoples and these are now called Roman, because the Romans use them very skilfully; in addition they took military exercises from others.
>
> Arrian, *Tactica* 33.2-3

Rome's army was a flexible one. Its overall organization, its equipment and its fighting techniques changed as Rome encountered new enemies, new topography and new military problems. Military units could be increased in size in times of emergency, a much less disruptive way of increasing the numbers of troops than raising new units from scratch. Alternatively, once the army was established on a permanent footing in the early Principate, units might be retained under or over their nominal strength as the military situation demanded. It is therefore not at all surprising to find considerable discrepancies between the theoretical 'paper' strength of units and their actual strength, and indeed we should expect to see such a degree of variation. The theoretical work of Pseudo-Hyginus has to assume that all units were of a standard size and type, but tacitly accepts that in reality this may not be the case: the author provides the space allocations not only for each type of unit, but for the individual legionary, auxiliary infantryman and cavalryman. This allowed the surveyor to calculate the space required for every unit in his army, even those vastly over-strength or under-strength, or with 'non-standard' organizations. The manipular and cohortal systems were very successful because of their great flexibility, and this included flexibility in the size and organization of units.

2. On the move: the army on the march

Introduction

> Those who have made a careful study of the art of war assert that there are usually more dangers on the line of march than even in the line of battle. For in pitched battle the men are armed, they can see the enemy approaching and they come mentally prepared for fighting; on the march though the soldier is not as ready with his weapons, not as attentive, and a sudden attack or treacherous ambush quickly confuses him.
>
> Vegetius 3.6

The first part of Vegetius' statement might initially appear to be excessively alarmist, but his elaboration of the topic justifies his concern about the vulnerability of the marching formation: dangers are present because the soldiers are not prepared to fight, either mentally or physically, and can therefore be easily taken by surprise. His concerns are more understandable when one considers that Roman armies experienced defeats when in marching formation, that were as serious as those encountered in set-piece battles. Trasimene in 217 BC was one such defeat, and the Varian disaster in AD 9. The loss of three legions with Varus had a profound effect on Roman frontier policy, and Augustus is reported to have beaten his head on the door demanding Varus to 'Give me back my legions!' A marching column that was in retreat was in particular danger, the morale of its troops likely to be low, the column less likely to keep to proper order, and the enemy encouraged to attack. The majority of especially serious defeats of marching columns occurred during a retreat. The marching formation, as Vegetius says, was vulnerable to ambushes, but particularly when proceeding through difficult terrain such as hills or forests, and the line could easily be thrown into disorder. If the column was too long or unprepared, it could be all but wiped out, as Dio's graphic account of the destruction of Varus' legions in Germany shows.[32]

The Varian disaster

Varus had led his three legions out on an expedition in the winter and had been drawn into an ambush by the German leader Arminius. The Roman soldiers were marching through forests, which required the advance guard to clear the route, but they were accompanied

8 *An army marching out, accompanied by standard bearers and horn players, Trajan's Column.* Courtesy JCN Coulston

by large numbers of wagons, pack animals and servants, and even by women and children. All of these factors caused the line to become strung out in scattered groups. Dio describes the formation as being more appropriate for peacetime, which the unhappy Varus had been lured into believing was the case. The situation worsened with a heavy downpour that caused the column to become even more fragmented.

> As the Romans were struggling against the elements, the barbarians, knowing the paths through the densest thickets suddenly emerged on all sides and surrounded them. At first, they hurled their missiles from a distance but then, as the Romans failed to fight back and many were wounded, they ventured closer. The Romans were not marching in any regular order but were mixed up with the wagons and non-combatants, so were unable to form up properly; they were outnumbered at every quarter, suffered high casualties and were unable to counter-attack.
>
> Dio 56.20

After several days of increasingly difficult progress and growing casualties, Varus and his officers lost hope. Velleius Paterculus mentions that some officers abandoned the infantry and attempted to flee with the cavalry, whilst others, including Varus himself, committed suicide. Many of their soldiers followed this example or were killed by the Germans. The whole column was wiped out; only a few escaped or were captured. Tacitus describes how Germanicus discovered the site of the disaster six years later:

> They were not far from the Teutoburg forest in which the remains of Varus and his legions were said to be lying unburied.
> Germanicus was filled with a desire to pay his last respects to the soldiers and their general, and every man with him was moved to pity for his relatives and friends, and they reflected on the dangers of war and on human fortune. Caecina was sent ahead to scout the secret woods and to build bridges and causeways on the soggy marshland with their lethal surfaces; then they advanced to the miserable places and the reality lived up to expectations. Varus' first camp with its broad dimensions and headquarters marked out was testament to the work of the three legions. Then a half-ruined rampart and shallow ditch indicated where the last survivors had huddled. On the open ground were the whitening bones, scattered where they had fled, heaped up where they had rallied. Bits of weapons and horses' limbs lay about, and human heads fixed to tree-trunks. In groves nearby were barbaric altars, where the Germans had slaughtered the tribunes and senior centurions. Survivors of the disaster, who had escaped from the battle or captivity, showed where the legates had fallen, where the eagles had been captured, where Varus had received his first wound, and where he had died by a blow from his own miserable hand. And so six years after the disaster a Roman army had come to bury the bones of the men of three legions.
>
> Tacitus, *Annals* 1.60-61

In 1987 the site of the disaster was discovered again, this time by chance, and fully investigated. Finds of Roman military equipment, artefacts and coins dating to exactly the right period identified the site as that of Varus' defeat, at Kalkriese a few miles north-east of Osnabrück in Germany. The scatters of finds show that Varus and his army were caught on sloping ground between the hill of Kalkriese to the left and marshland to the right. A heavy accumulation of finds probably indicates the area of the main attack and an attempt by the Romans to make a stand, but they suffered heavy losses. Then the survivors headed off in different directions, one group to the north west, another to the south west, both groups continuing to come under attack. Additional finds in the marshes may represent some Romans trying to escape in that direction, such as one of the standard-bearers whom Florus claimed escaped with his eagle by hiding himself in the marsh. The whole site provides a unique and still graphic picture of the destruction of an army **(10)**.[33]

Preliminaries

Soldiers spent a lot more time on the march than they did in pitched battle and the considerable importance of a properly ordered and disciplined marching formation is shown by the military regulations or *constitutiones* of Augustus and Hadrian. These required both infantry and cavalry to carry out route-marches regularly, and Vegetius claims that recruits frequently carried a pack of up to 60 Roman pounds for these practice

9 *Tombstone of M.Caelius, centurion of Legion XVIII, a casualty of the Varian disaster.* Courtesy Landesmuseum, Bonn

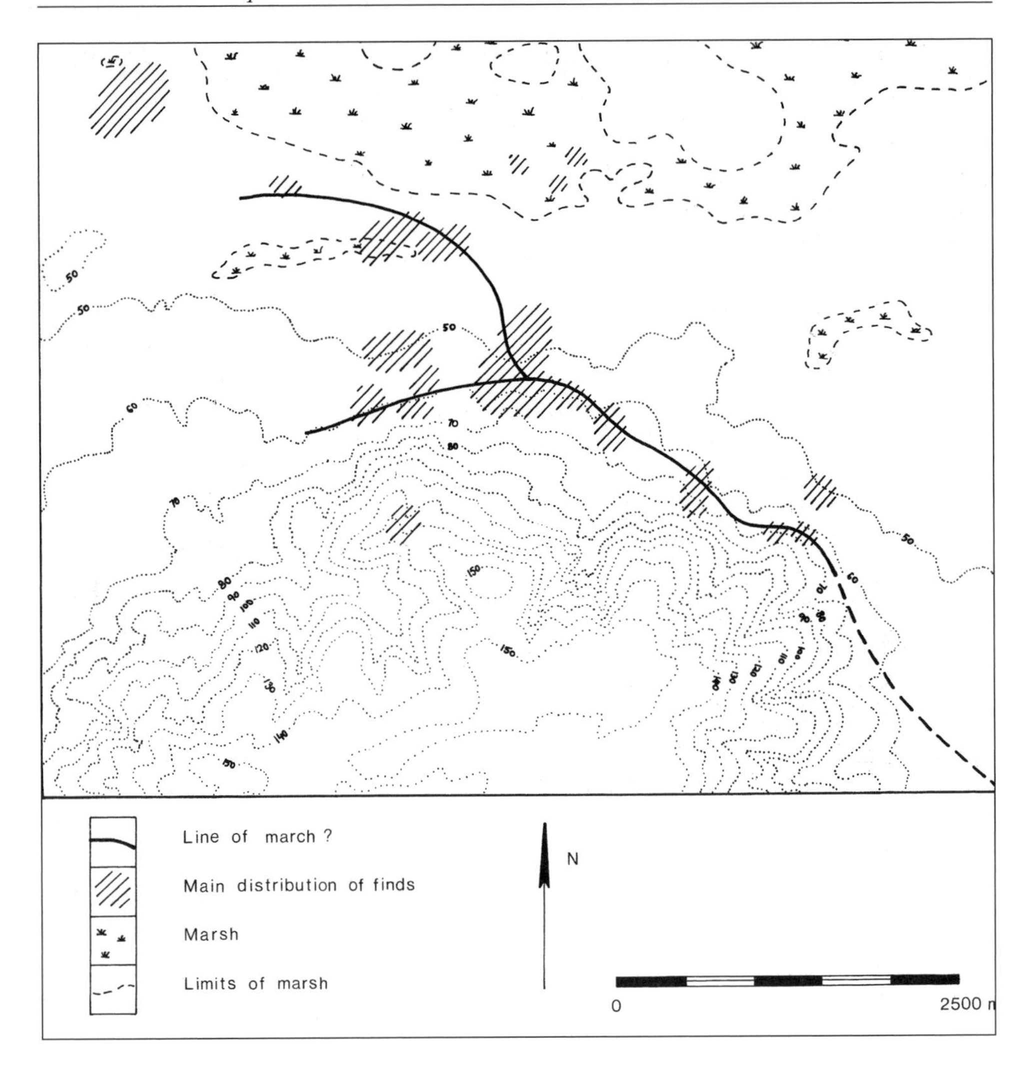

10 *The remains of the Varian disaster at Kalkriese, Onsabrück, Germany. The army was marching a north-westerly direction, between the steep hills of Kalkriese and the marsh, a perfect spot for an ambush (after Berger* et al.*)*

marches. Infantry had to march for twenty miles fully equipped, some of that distance at speed, whilst the cavalry covered the same distance but undertook manoeuvres en route. These route-marches were supposed to take place three times a month. Onasander also mentions the importance of such exercises, to familiarize the soldiers with marching in formation on the alert. We learn of a prohibition on falling out of the marching column, which may have been another military regulation. This was probably partly to ensure that soldiers did not sneak off to do a bit of plundering, but it is also likely to have been to maintain proper discipline and formation.[34] Historians frequently show how commanders attempted to keep control of their men on the march:

> Metellus moved up and down the column to see that no one left the ranks, that the men kept close together round their standards, and that each soldier carried his food and arms.
>
> [His successor, Marius] went round every section of the army (on the march) distributing praise and reprimands as they were deserved.

Sulla, Agricola and Trajan are also praised for similar actions whilst Vespasian was accustomed to marching at the head of the column. Arrian proposed to ride up and down his column, partly to ensure that the men did not fall out. Clearly riding along the length of the marching column was considered to be one of the actions of a good commander and indicates how much importance was placed on the maintenance of correct, disciplined formation. Soldiers focused on their standards when they formed up into ranks for pitched battle, and these would be of particular importance in the very vulnerable transition from one formation to the other. If the troops were properly ordered in the marching column around their standards, the manoeuvre into line of battle might be slightly less hazardous.[35]

The Varian disaster illustrates well the dangers of a poorly managed column and the considerable importance of a properly ordered marching formation. It is hardly surprising therefore, to find this subject given some prominence in the military handbooks. However, historians tend not to devote very much time to describing marching orders. Such descriptions rarely provided the historian with the opportunity to exercise his dramatic abilities in the way that a siege or set-piece battle could, unless of course a major disaster was unfolding, though there can be a certain impressiveness about a detailed description of an army on the march. Tacitus' account of the march into Rome by Vitellius' troops in AD 69 is colourful, and even more impressive are Josephus' descriptions of the armies of Vespasian and Titus in Judaea. In his witty tract on the writing of history, the second century AD Greek writer Lucian noted that descriptions of marching formations were a minor but integral part of historical narrative and accuracy was important. Despite this, most Roman historians' descriptions of armies' marching formations are frequently little more than passing comments. One document, which is of unique importance though in our understanding of this subject, is Arrian's *Order of March against the Alans* (*ektaxis*). Whilst governor of the province of Cappadocia under Hadrian, Arrian was faced with a possible invasion by the semi-nomadic Alans. He successfully dealt with the threat, probably without the need for a pitched battle, but he nonetheless published a report outlining his marching order and proposed deployments should pitched battle occur. The document provides valuable information on the relationship between line of march and line of battle, and we can suggest how the army might have manoeuvred from one to the other, a procedure fraught with danger because of the potential for confusion and an enemy attack.[36]

If the historical writers are brief in their descriptions of marching columns, however, they provide even less information on how the army provided for itself when on campaign, and on what logistical organization existed. They occasionally mention these in passing, but there are no extended treatments of the Roman army's system of supply and logistics in the way that Polybius and Josephus describe the methods of encampment and order of march, or the fourth-century historian Ammianus Marcellinus explains the various pieces of siege equipment. Nor, indeed, do the military treatises provide much information on logistics other than showing an awareness of the importance of securing supply lines and ensuring the army is properly fed. This may be because they, like the historical writers, tend to concentrate more on the actual waging of war than on the preparations for it. Such matters though affected the size and speed of the marching army, and the distances it could travel.

The organization of the marching column

Caesar reports an interesting episode concerning his army on the march in Gaul and an attempted attack on it. The Nervii had been informed by Belgic spies that each of Caesar's legions on the march was separated from the next by a part of the baggage train; this would make it easy for the Nervii to attack the first legion when it reached the campsite and was not yet supported by the other legions behind. The Nervii therefore prepared an ambush for the Roman troops. When Caesar arrived, the Nervii must have been surprised to see that his army was not formed up in this way at all. Instead, Caesar was marching at the head of a column with six legions in light order. The baggage train followed under the protection of two newly recruited legions **(11)**. Although the ambush caused the Romans difficulties, the legions were quickly able to form up a makeshift battle line and defeat the Nervii. Caesar reveals that the Belgic spies had not given their allies incorrect information, just reported the organization of his normal marching formation. However, when the Nervii encountered him he was employing his normal marching formation *for use when approaching the enemy*.[37]

As noted above, in his account of the Varian disaster Dio censures Varus for advancing with his army in a formation that was more appropriate to peacetime than war. Conversely, Sallust twice mentions that the Roman army in the war against Jugurtha was marching through relatively secure territory, but the column was arranged 'as if the enemy were close at hand'. The commanders of the armies were Metellus on the first occasion and Marius on the second, both men whom we have already noted took care to ensure that their armies were well disciplined on the march. Under some circumstances, a Roman army was expected by contemporary writers to adopt a specific marching formation that was 'ready for battle, in a place where the enemy might appear from any quarter', but this would only have been necessary if attack was likely. It is hardly surprising to find that the Roman army's marching formation varied to take into account specific circumstances such as topographical factors, and particularly the tactical situation, the likelihood or not of being attacked or having to fight a pitched battle with the requirement to redeploy from line of march to line of battle. This is why the Nervii were caught out by the change in Caesar's marching formation. A look at the ancient textbooks and other descriptions of marching columns shows that the Romans employed several different basic formations, though the details tended to vary according to the immediate situation and the preferences of individual generals.[38]

The earliest detailed description of a Roman army's marching formation is that given

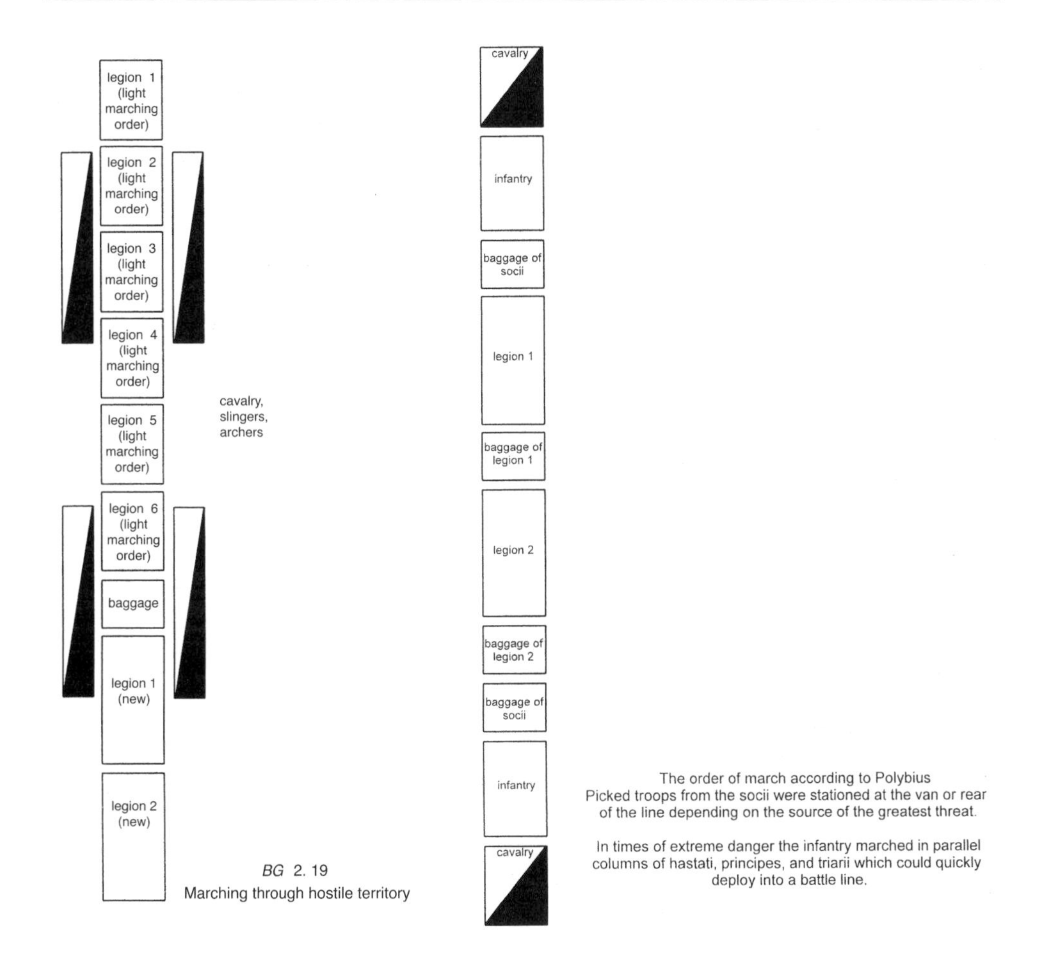

11 *Caesar's order of march against the Belgae. (Not to scale)*
12 *The order of march according to Polybius. (Not to scale)*

by Polybius as part of his excursus on the Roman military system. The baggage of the various sections of the army is spread out along the column to avoid a long baggage train dividing up the army too much **(12)**. With a small army, (the consular army described by Polybius contained only two legions), it was important that the troops remained in contact with each other. Polybius states that the Romans used two different marching formations, the one described above for use under normal circumstances, the second being employed in times of extreme danger when there was the probability of attack. It is under the latter circumstances that one would expect to see a close correlation between line of march and line of battle, to shorten the dangerous period of re-deployment. Polybius' second marching formation is essentially the battle line: a simple left or right turn would convert it from one to the other. This might sound straightforward, but with the manipular legion of this period the lines of the battle formation were not uniformly armed and equipped.

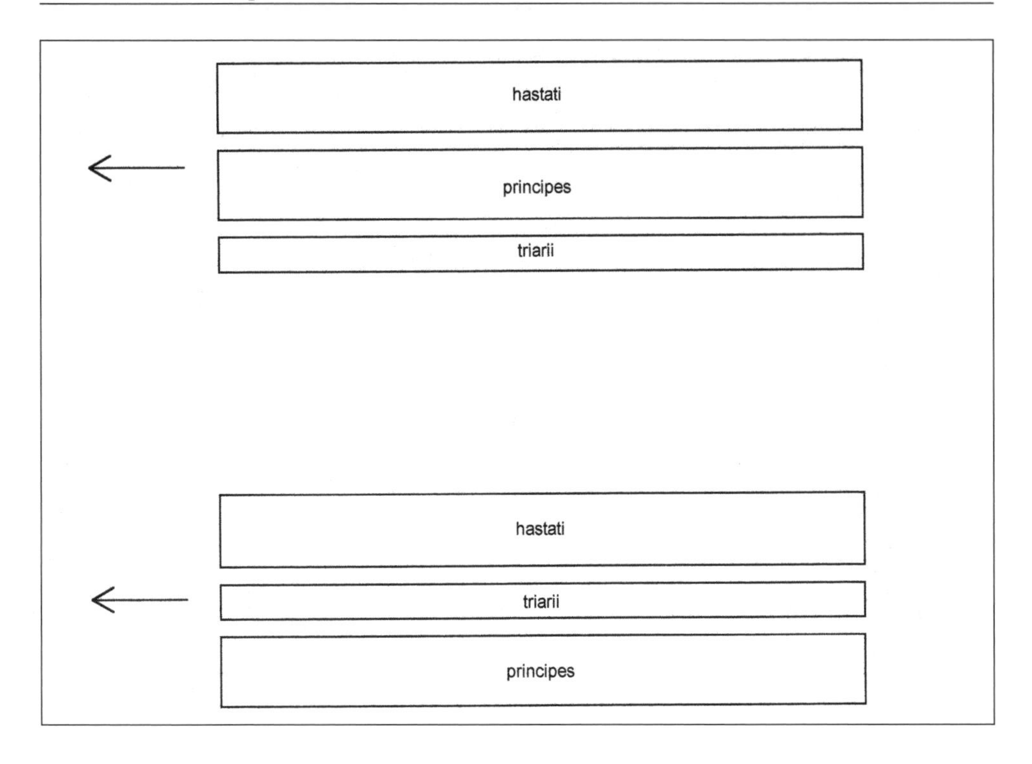

13 *The 'Polybian' legion on the march, above, and Walbank's suggestion which ensures the* hastati *or* principes *meet an attack*

The simple turn into battle line could theoretically result in the *triarii* facing the oncoming enemy rather than the *hastati*. This has led Walbank to speculate that when the legion was marching in hostile territory, the *triarii* would march in the middle of the column, flanked by the *hastati* and *principes* **(13)**.[39] Then if the army had to form up suddenly into battle line the *triarii* could manoeuvre to the rear of the line, leaving either the *hastati* or *principes* to face the attack, both of which groups were armed and equipped for the first encounter of pitched battle. This solution would have reduced the amount of manoeuvring necessary in case of such redeployment. The cohortal legion, however, faced no such difficulties since all the soldiers were uniformly armed.

The recommendations of Onasander and Vegetius are fairly similar to those of Polybius, though Onasander does not give as much detail on what kind of troops are placed at particular points in the line. Onasander and Vegetius both mention the use of scouts, and have only one baggage train, but this is flanked by troops for added protection **(14)**. All three authors note that picked troops were stationed to provide a protective screen to the section of the column that was likely to come under the greatest threat, which would allow the main body of infantry time to react and prepare themselves for an attack.

If we turn to descriptions of marching formations provided by historians, we can see that in general they are similar to the formations advised by the theoreticians **(15)**. In secure territory, usually the legions marched in the centre of the column with the allies or

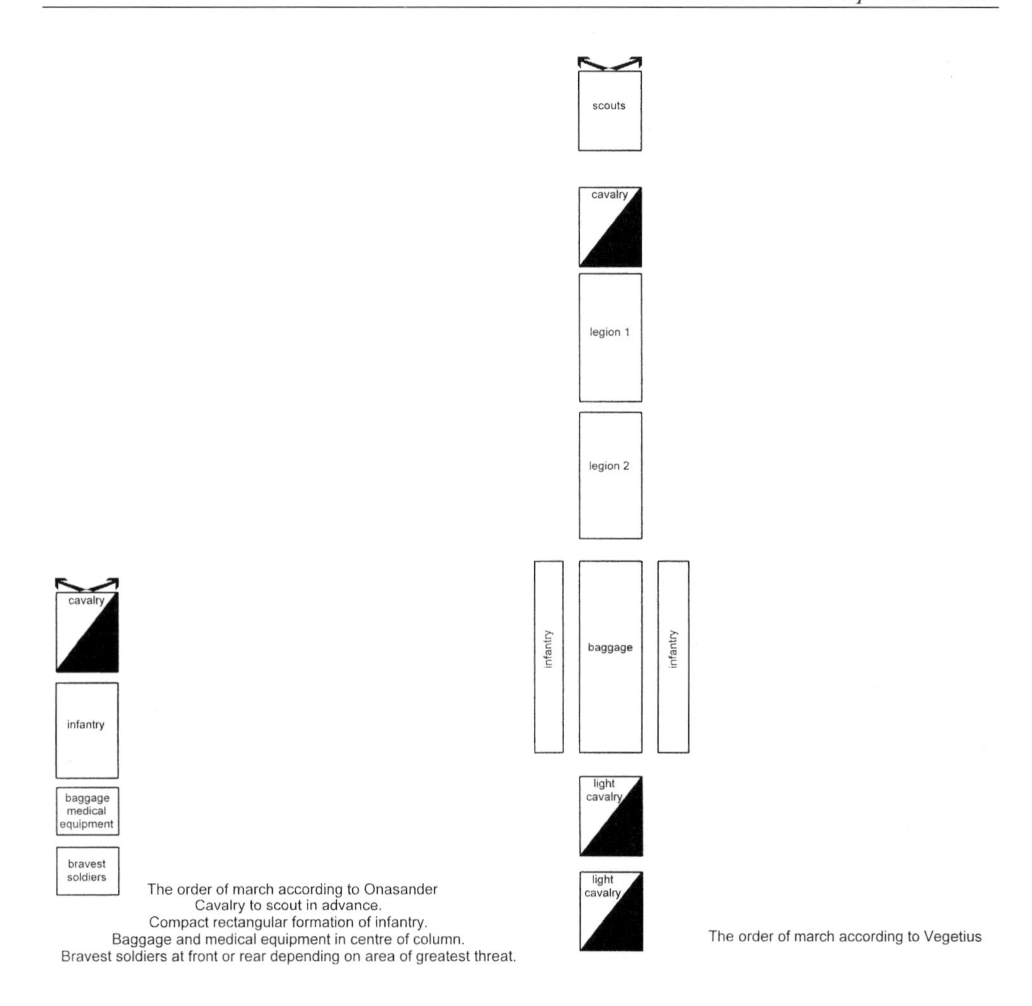

14 *The order of march according to Onasander and Vegetius. (Not to scale)*

auxiliaries at the front and rear. Cavalry were normally positioned as van and rearguards. The general and his retinue, including bodyguard and in the Empire the Praetorian Guard if the emperor or a member of the imperial family was on campaign, were in the centre of the column, before or between the legions. The baggage train and any siege equipment was also usually placed safely in the centre of the column, or it could be divided up if it was particularly large, or if the column was likely to be attacked, as Caesar did when marching against the Nervii. In Josephus' two descriptions, the general is positioned before the main legionary force whilst the officers' baggage, siege equipment, and the main baggage train are in separate sections of the column **(16, 17)**. Caesar, on the other hand, regularly placed his baggage train towards the rear of his column where they were unlikely to be attacked, and these were followed by his less experienced legions. Flank guards were not always deployed, and may only have been used when there was the danger of attack, or if the general intended to manoeuvre into line of battle under a

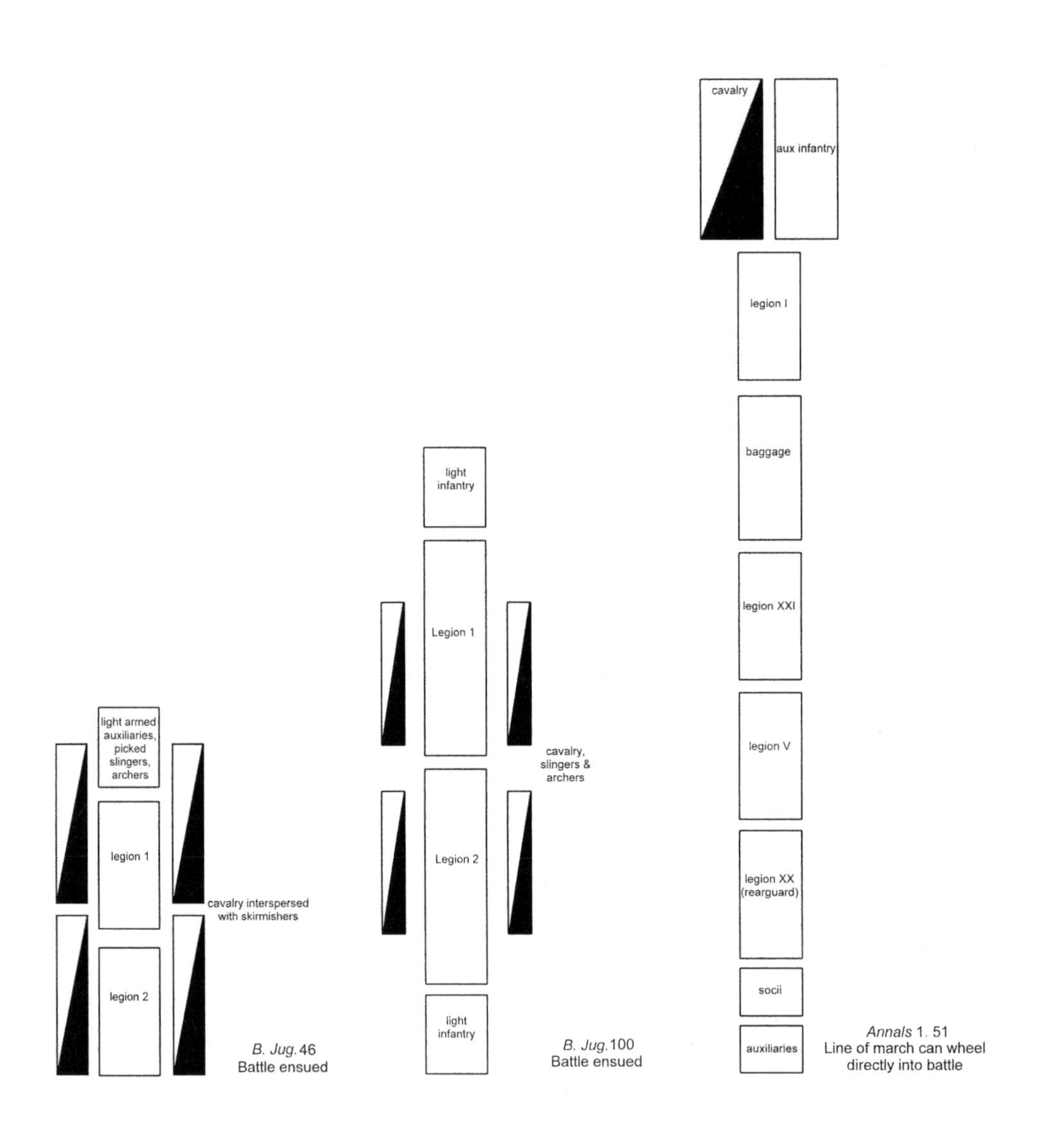

15 *Orders of march from 100 BC to AD 58. (Not to scale)*

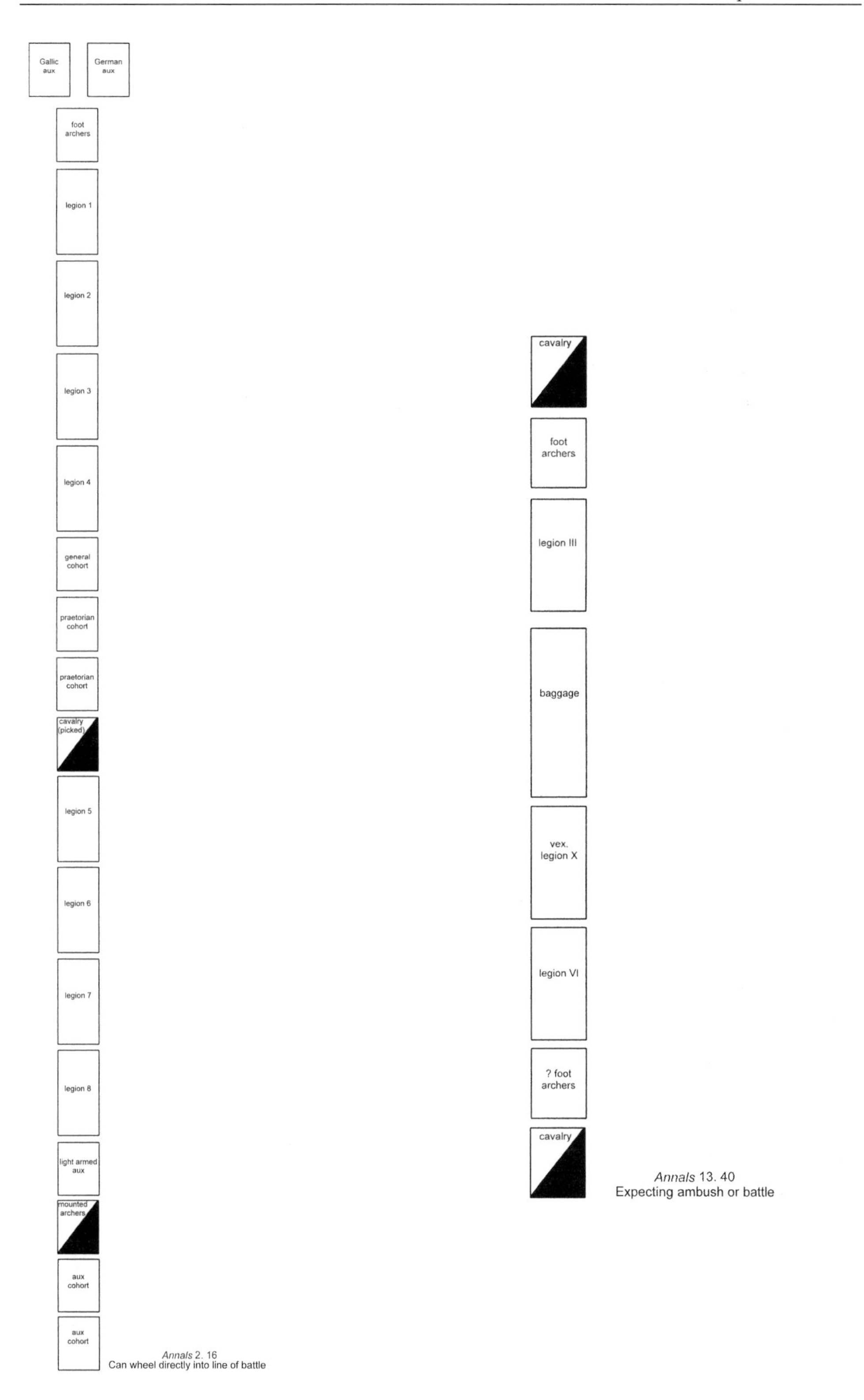

Annals 2. 16
Can wheel directly into line of battle

Annals 13. 40
Expecting ambush or battle

protective screen of cavalry. The order of legions within the marching column probably also varied. Polybius states that the order varied to allow all ranks an equal opportunity to find a fresh water supply and fresh foraging ground. Josephus though claims that lots were drawn for the legion that was to head the column (and would therefore have had first opportunity to obtain fresh water and forage). By placing his inexperienced legions at the rear of his column, Caesar ensured that they were likely to be kept further away from enemy attacks, which could be dealt with by the more experienced legions. At the rear of the column though, they would have had to content themselves with fouler water supplies and, positioned behind the baggage train, would have been toughened up by having to march through waste from the baggage animals.[40]

Historical accounts frequently note the deployment of scouts in advance of a marching column. Scouts would have gathered intelligence on the nature of the topography, location of supplies, suitable places to camp, and the location and nature of the enemy. The cavalry in the van of Vespasian's army would probably have fanned out in advance to scout the terrain, and Arrian used his cavalry to scout ahead of his marching column. In Arrian's case good intelligence was particularly important since he was campaigning against a highly mobile enemy and was intending to deploy his battle line under specific topographical circumstances. Caesar's scouts, or *exploratores*, are reported as gathering intelligence up to a day's forced march away. In addition to scouts, engineers and surveyors were sometimes sent in advance of the main marching column to clear obstacles from the route and survey the campsite. Josephus reports that the soldiers Vespasian sent ahead of his army along the difficult mountain route from Gabora to Jotapata transformed it in four days into a broad highway suitable for his heavy infantry and siege train. The theoretical army described by Pseudo-Hyginus included 1300 marines from the fleets at Misenum and Ravenna; they all camped in the *praetentura* because they led the army out of the camp and constructed roads, presumably also clearing obstacles.[41]

Various attempts have been made by historians to analyze the different types of marching formation, categorize them and provide explanations for their use under different circumstances. Le Bohec sees two main types of marching column during the imperial period and suggests that the use of these was dependent upon topographical circumstances. He argues that the narrow column without the flank guards used by Caesar and Titus was for use in confined terrain whereas the column with flank guards, as used by Germanicus and Arrian, was for level, open terrain. Obviously, there had to be some relationship between the arrangement of the line of march and the kind of terrain through which the army was marching. A Byzantine treatise explains in some detail how the width of the column should be decreased and subsequently increased again as confined areas are negotiated, but the column returned to its normal arrangement afterwards.[42]

Although topographical considerations were important, all the ancient literary evidence points overwhelmingly to the fact that the arrangement of the line of march depended principally on the tactical situation. If the column was marching through allied territory, or there was no chance of an attack, a long narrow column might be deployed, as described by Polybius and used by Caesar. This formation was also employed by Vespasian and Titus in Judaea **(16, 17)**. Josephus does not mention any flank guards on these lines of march and

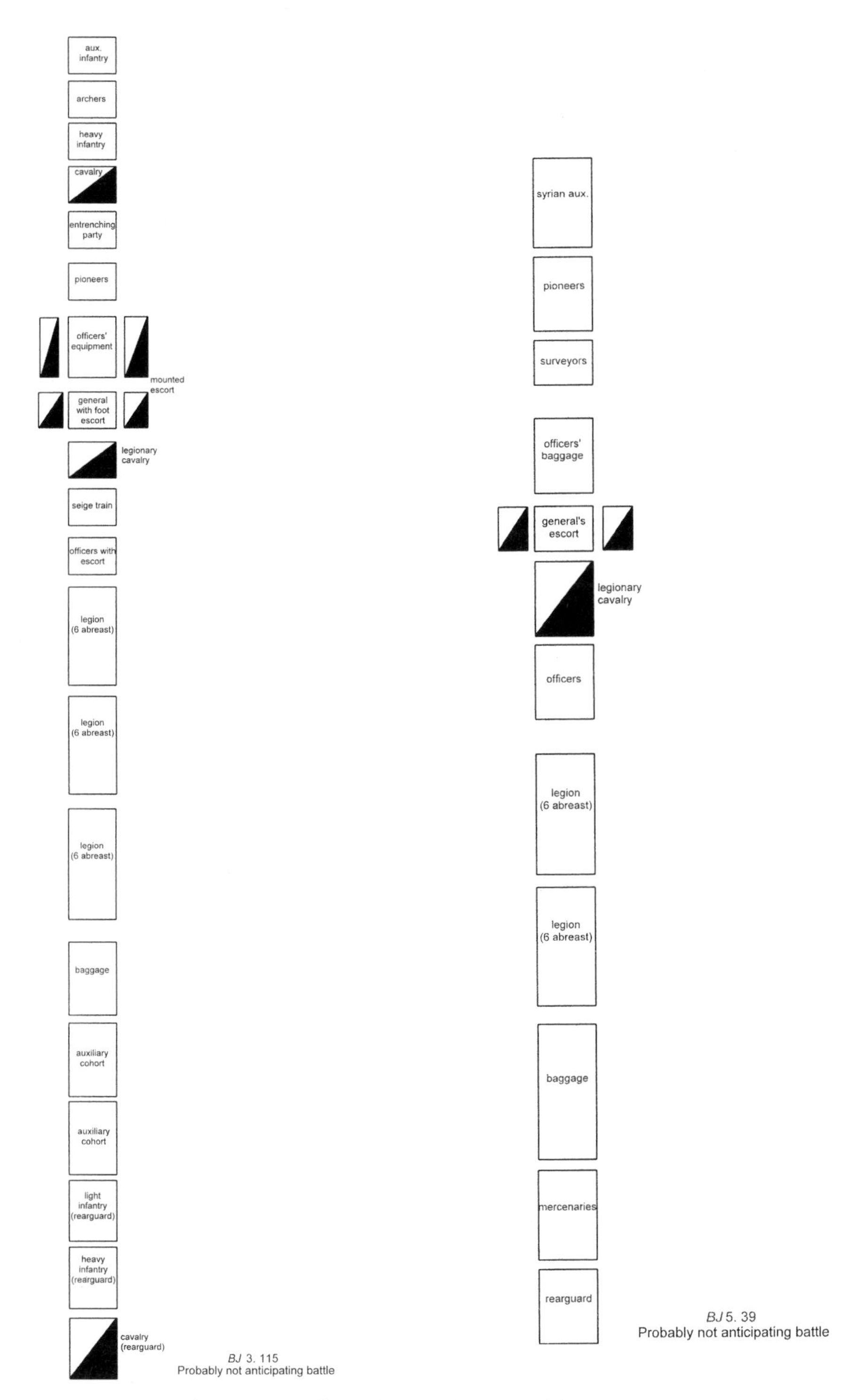

16 *Vespasian's advance into Judaea, AD 67. (Not to scale)*

17 *Titus' march on Jerusalem, AD 70. (Not to scale)*

it is unlikely that Vespasian would have deployed his column in this way if he had been expecting an attack. Caesar criticizes Sabinus and Cotta for allowing a long, straggling marching column to leave the relative safety of their camp in 54 BC when their army was trying to escape from Gallic rebels. He claims that Sabinus had not anticipated an attack and when it came, he had to rearrange his cohorts in a hurry. The implication is that had Sabinus been expecting an attack, his marching formation would have been different and he would not have been forced into sudden manoeuvres which panicked his men.[43]

The type of marching formation for use in insecure circumstances was the *triplex acies* (triple battle line) or *agmen quadratum* (squared column). Its arrangement would have been closely related to any subsequent line of battle and, as Polybius says, it could turn or wheel to face an enemy in pitched battle. The very dense and compact nature of this formation is illustrated by Livy who describes how soldiers at a military show assembled in this fashion in preparation for forming the *testudo*, another tightly packed formation. Soldiers may have been trained specifically to manoeuvre from *agmen quadratum* to *testudo* and line of battle. The *agmen quadratum* was both a defensive and an aggressive formation: it provided protection in case of retreat, but it was also used when the Romans were actively seeking battle. Because of its compact organization, soldiers were less likely to be isolated by enemy cavalry or archers, and the large shields of the legionaries would have provided extra protection. When historians describe this type of marching formation they frequently comment on the column's ability to wheel directly into line of battle from marching formation. In many of these cases, the army did form up to fight a pitched battle, or the commander expected to do so, or feared an attack on his column. Corbulo's line of march in Armenia was his battle line, formed up in this fashion because the general was expecting to encounter an ambush or pitched battle. Because of the danger of ambushes, particularly in hilly or wooded country, as Varus found to his cost, the topography might influence the general's choice of formation. Hence, Germanicus tended to employ a *triplex acies* formation in his campaigns in the forests of Germany, and in Africa when marching through wooded territory Metellus also deployed his army in a close formation with flank guards.[44]

Length and width of marching columns

A long narrow line of march was in particular danger in hostile territory and as discussed above, generals were advised to use a more protective compact formation under such circumstances. No particular column width is recommended by the treatises, and the actual width in men may not have been of great importance provided that the column was wide enough to prevent it being pierced by a flank attack, or to produce a battle line of sufficient depth if it wheeled into this formation. Onasander advises against a long extended column for several reasons, including a concern that such a formation was more likely to induce panic and apprehension due to uncertainty, and includes a scenario illustrating what might happen:

> Sometimes the van, after descending from hilly terrain into open, level land, on seeing the rearguard still descending behind them have mistaken them for an

> attacking enemy, with the result that they intended to march on their own men as enemies, and some have even come to blows.
>
> Onasander 6

Such a misidentification could really happen, and caused panic in Vitellius' army in AD 68. Most of the army had settled for the night and was alarmed by a distant cloud of dust and glint of weapons indicating an advancing army. The situation was calmed when it was realized that the arriving troops were not the enemy but the rearguard of Vitellius' own army, arriving at camp. Elsewhere Tacitus states that a column with a long baggage train was easy to ambush and difficult to defend. These dangers are well illustrated in the historical literature. We have seen how vulnerable the ill-disciplined, straggling column of Varus was, and Caesar mentions that the retreating column of Sabinus and Cotta in 54 BC was too long to be properly controlled, and this contributed to the ensuing panic and disorder among the ranks.[45]

In depth analysis of the actual lines of march is difficult because there are so few sufficiently detailed descriptions surviving. An attempt to calculate the length of Vespasian's line of march in Judaea, as described by Josephus, resulted in a marching column some 17-18.5 miles (28-30 km) long. So long, indeed, was the column, that 'the head of the troops entered camp for their overnight rest before the last of the troops were able to leave the site of the previous overnight stay'.[46] Although the organization of the column and absence of flank guards indicates that Vespasian expected no attacks on his army, such a line would have been extremely vulnerable. Whilst it might have been necessary for the army to extend itself considerably to negotiate some of the more mountainous regions, it seems unlikely that the line would have been this long at all times, or indeed ever. The army only had a core of three legions. Armies may well have marched in parallel columns whenever possible to reduce the length of the marching column overall. Because the campaigning season coincided with the best weather of the year, in some terrains this may not have posed major difficulties, but in poor weather or mountainous or marshy terrain the army may have had no option to follow defined roads or tracks, and employ a long and vulnerable formation.

The only other description of a line of march that survives in sufficient detail for this kind of analysis is Arrian's order of march against the Alans. This is a far more detailed account than Josephus' descriptions, in that Arrian lists the individual units in his army, but even here there are serious problems in calculating its length. Most of Arrian's units can be firmly identified but we do not always know whether the units were *quingenary* or *milliary*. In addition, some of the units were, we know, present only as vexillations or detachments, and since part of the Cappadocian army was in Judaea at this time, the others may not have been at full strength either.[47] Finally, although Arrian mentions that the legions marched four abreast, he does not indicate how many abreast the auxiliaries or cavalry marched. These calculations give a column length of about 3.5 miles (4.8 km), compact enough that communications along the line should not have been too difficult and that redeployment into line of battle should not have taken an excessive length of time.

Hypothetical reconstruction of Arrian's line of march (**21**)

Because of the problems involved in this excercise, these calculations contain the following estimations and assumptions:
i) All regular army units are assumed to be at full paper strength (unless known to be present only as vexillation), and the vexillation of Legion XII Fulminata 2000 strong. The provincial militia is estimated at c.800. Because Arrian is marching to pitched battle and no camp is to be constructed before the battle, it is assumed that the bulk of the army's baggage remains in a previously constructed guarded camp.
ii) All infantry march in columns four abreast, the cavalry three abreast.
iii) 0.75m is assigned to each infantry rank, 2.7m to cavalry ranks following 19th century handbooks.
iv) 100 metres are added to provide gaps of at least 5 metres between the units.

UNIT	NUMBER	LENGTH (metres)
Scouts	In advance of main column (not included in calculations)	
Cavalry		
Coh III Ulp. Petraeorum sag. ∞ eq.	240	216
Ala II Ulpia Auriana	512	460
Coh IV Raetorum eq.	120	108
Ala I Aug. Gemina Colonorum	512	460
Coh I Ituraeorum eq.	120	108
Coh I Aug. Cyrenaica. eq.	120	108
Coh I Germanorum eq.	240	216
	1984	1784
Infantry		
Coh I Italica	480	90
Coh I Aug. Cyrenaica eq	480	90
Coh I Bosporanorum ∞	800	150
Coh I Flavia Numidarum sag.	480	90
	2240	420
Equites Singulares	120	108
Equites Legionis	180	162
Artillery		100
Officers Legion XV		50
Legion XV Apollinaris	4800	900
Officers Legion XII		30
Legion XII Fulminata	2000	375
	7100 +	1725
Infantry		
Provincial militia	800	150
Coh I Apula	200	38
Impedimenta		200
	1000 +	388
Cavalry		
Ala I Ulpia Dacorum	512	460
Gaps between units		100
	12836	4877
		c.3.5 miles

Marching distances

> When he was campaigning in Spain, Marcus Cato realized that he could capture a particular town if he took the enemy in a surprise attack. So having completed a four day's march in two days through rough and barren terrain he defeated the enemy who had not feared an attack. When his troops asked the reason for such an easy victory, he told them that they had won the victory as soon as they had completed the four days' march in two.
>
> Frontinus, *Stratagems* 3.1.2

Unfortunately, there is very little good evidence for the distances that a Roman army might march in a day and the average speed of an army. Like the organization of the column itself, speed and distance would have been affected by the nature of the terrain, the tactical and strategic situation and the composition of the army. A force with a long baggage train, heavy siege equipment and many non-combatants would inevitably have moved more slowly than one that was comparatively lightly equipped. Commanders were generally reluctant for their soldiers to march long distances and then immediately fight a pitched battle, so the proximity of the enemy would also influence how far an army marched in a day.

As part of their training exercises, soldiers were supposed to complete a five-hour march of 20 miles (1 Roman mile = 1618 yards) at the normal military step, but 24 miles at the full step. These timings were for soldiers on exercise, not for a campaign army with a baggage train having to deal with the different factors mentioned above. Nonetheless, they may give some idea of the distances an army could be asked to cover under optimum conditions, and such distances were not unrealistic. Caesar, renowned for his *celeritas*, the rapidity with which he could move with his forces, got his army to march 25 miles from camp to intercept the Aedui during the Gallic revolt. Then, after just three hours rest during the night, he proceeded towards Gergovia, but had to double back when his original camp was attacked. His soldiers covered over 50 miles at speed within a couple of days. Whether they would have been in much of a state to fight after these forced marches is another matter. In another military emergency, Caesar's soldiers had to march 20 miles in a single day to the rescue of Quintus Cicero, but he could not leave his camp until Crassus arrived with reinforcements to occupy Caesar's main base at Amiens. Crassus left his camp 25 miles away at midnight, and by the third hour (about 9 am) his advance guards had arrived. This dashing around was done by unencumbered armies in emergencies or, in the case of Cato in Spain, to take the enemy by surprise. Probably the distances and times are reported precisely because they are exceptional. Much of the time armies probably moved much more slowly.[48]

Archaeological evidence can provide some information on the subject of march distances. Groups of marching camps in Scotland have been tentatively identified on the grounds of similar size, design, and style of defences. The dating of marching camps is notoriously difficult, but two distinct groups of camps have been traced in north east Scotland and dates suggested; one set has been attributed to Agricola's campaigns of the late first century AD, the other to Septimius Severus' early third century expedition. The

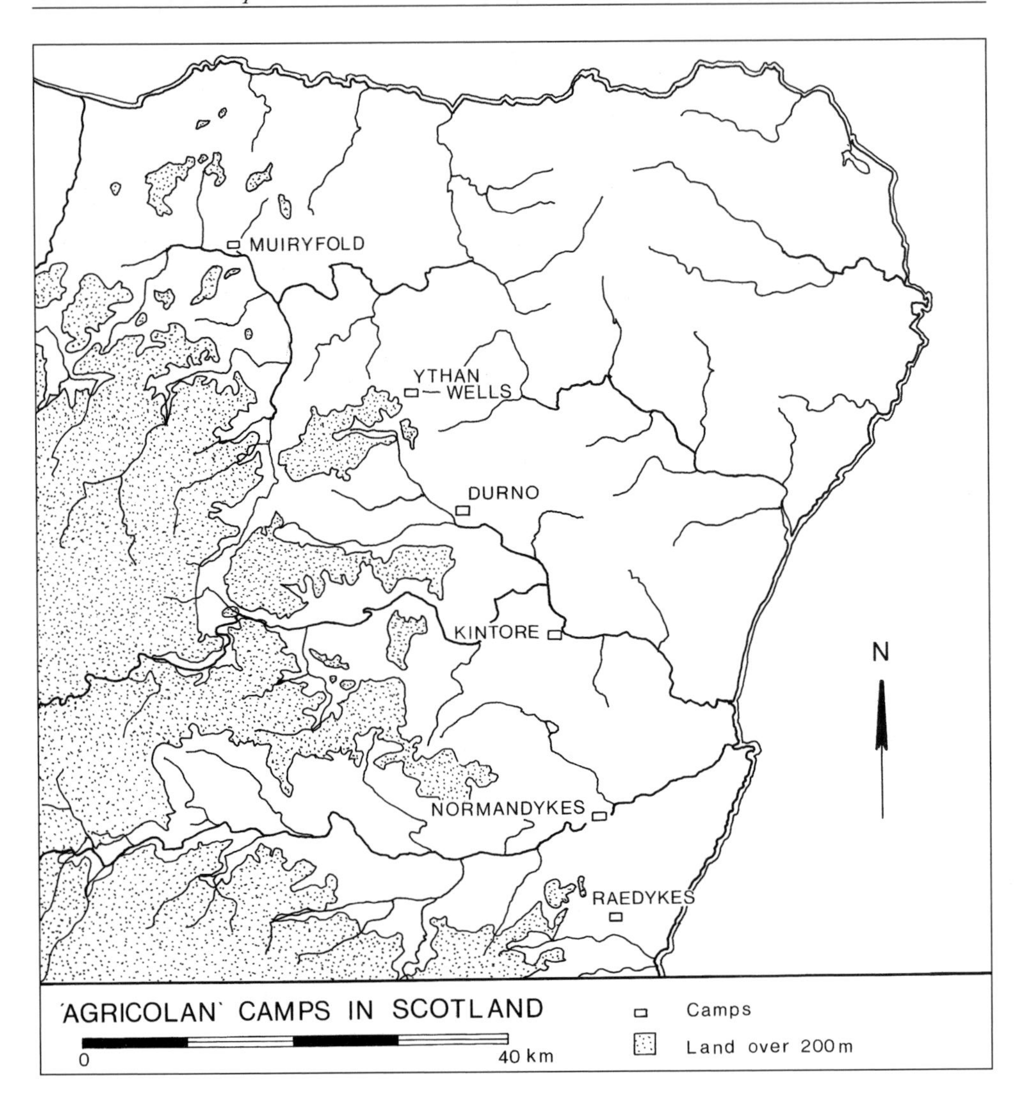

18 *'Agricolan' marching camps in Scotland (after St Joseph)*

distances between the camps in the two groups give some indication of the daily length of march of the armies on the campaigns **(18, 19)**. These armies then were covering only very short distances each day, not even 15 miles, and sometimes much less. The shortest day's march, between Raedykes and Normandykes, involved the army crossing the river Tay, which would have slowed progress down considerably. Neither Agricola, until his last campaign, nor Severus, had much luck in forcing encounters with the enemy, so there may have been no reason for either general to charge around at speed like Caesar, who had a more accessible enemy and clear centres of occupation to aim for. The groups of camps may then represent slower and more deliberate campaigns than many of Caesar's, but they do indicate very well the slowness with which many Roman armies must have moved.

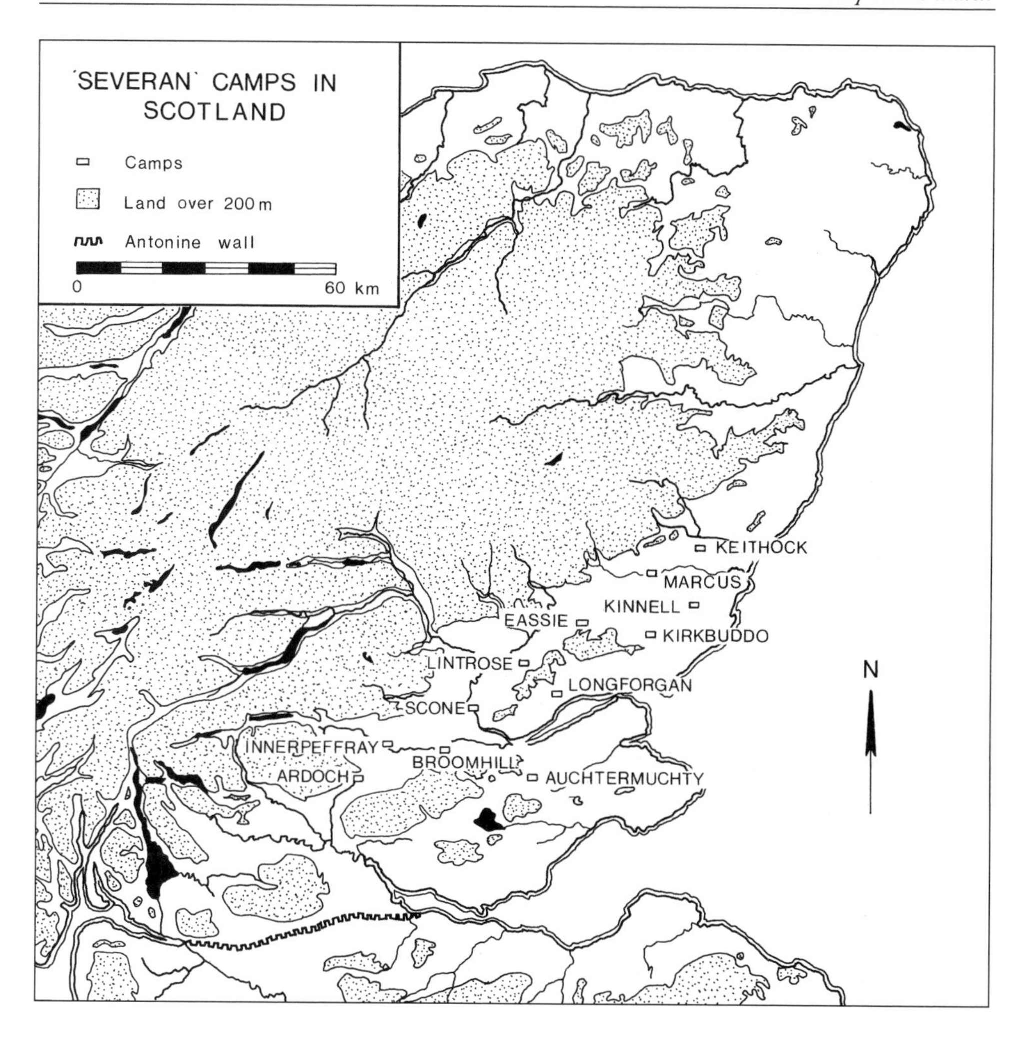

19 'Severan' marching camps in Scotland (after St Joseph)

Facing Attack

The vulnerability of an army on the march, even a well-organized and prepared army, is clearly illustrated by the regularity with which ancient armies, not just Roman armies, were attacked in these circumstances. A comparatively small force could disrupt the progress of a much larger army if it ambushed the marching column, as Caesar discovered in Africa in the civil war against Labienus. The latter's intention was to delay Caesar's march so he would be forced to camp where there was no water supply, and he used mobile light-armed troops to harry Caesar's column from the rear. Caesar was forced to halt and drive off the initial attack before continuing, and with repeated attacks by Labienus progress was very slow: the column alternately marched, then made a stand to drive off the enemy. Caesar did make his intended campsite, but a lot later than he had

20 *Two legionaries on the march, mid first century AD, Mainz.* Courtesy Landesmuseum Mainz

planned. He responded by altering his marching formation so that when Labienus next attacked, Caesar was prepared with a reserve force of three hundred men in light marching order who would not be fatigued from carrying heavy kit. Together with the cavalry, these legionaries drove off the enemy.[49]

When an army was attacked on the march, probably the most important factor in repelling the enemy was discipline. When Antony's column was ambushed by the Parthians and came under heavy missile fire, the soldiers formed themselves up into a protective *testudo* immediately without the need for elaborate orders from their officers. It is likely that the soldiers had been trained to react in just this way. Caesar's men too reformed their marching column automatically when it came under attack, halting and forming up in a hollow square with the baggage in the centre. Lack of discipline and training, on the other hand, contributed greatly to the disorder of the marching column led by Cotta and Sabinus. When ambushed by Ambiorix, the soldiers were ordered to abandon their baggage and form up into a tight defensive formation. But they were reluctant to leave their possessions and wasted time searching through their packs for valuables.[50]

Attacks on marching columns happened most frequently when the army was advancing without having deployed scouts. Failure to deploy scouts might seem to be an extraordinary omission on the part of a commander, but in spite of the advice of the treatise writers it happened surprisingly often. At Trasimene Flaminius marched his army of over 20,000 straight into Hannibal's ambush. He had made no attempt to reconnoitre the area despite the dawn mists and the proximity of the enemy; a couple of years previously Lucius Manlius had suffered heavy losses when marching through woodland without reconnaissance. It was not until his troops reached open ground that they could redeploy and make a stand, at which point his Gallic enemy departed. Whether or not Lucius Postumius sent scouts into the Litana forest is unknown, but the Boii caught the Romans by surprise by cutting into the trees so their trunks were almost severed. When the Romans entered the forest, the Boii toppled the trees onto the Romans, killing 25,000, including the general. It is hardly surprising that the ancient textbooks stress the importance of scouting in advance and determining the nature of the terrain as well as the enemy.[51]

Attacks on marching armies and their counter-attacks could very easily lead into pitched battle. The consul Scipio marched out to force an encounter with the pro-Carthaginian Indibilis in 212 BC, but when the two armies met there was no time to form battle lines so the encounter was more of a running battle between two marching formations. The Romans had the upper hand until their general was killed, whereupon they fled. The major battle at Cynoscephalae in Greece in 197 BC also began as a chance encounter between the advance guards of the Roman and Macedonian armies. The skirmishes eventually turned into a pitched battle with the complete destruction of Philip V's army. The suddenness with which a march could turn into battle explains why the treatise writers stress the importance of maintaining formation, discipline and alertness on the march.[52]

Redeployment: line of march to line of battle

All the ancient writers, both historians and theoreticians, stress the vulnerability of the line of march, and with good reason as we have seen. But an army was equally vulnerable to attack when it was redeploying from its marching formation to a line of battle, as Philip V found at Cynoscephalae. The soldiers were more likely to be concentrating on their manoeuvres than on the enemy, and a partially formed or broken battle line could be put to flight much more easily. It was probably for this reason that Lucullus, when badly outnumbered by Mithridates' army, ordered his troops to attack the Pontic forces whilst they were still not fully deployed. The size of the Pontic army meant deployment was very slow and it was unable to react in time, so Mithridates' soldiers simply turned and fled. Since deploying from line of march to line of battle was such a potentially dangerous time for any army, it was important that there was a close correlation between the two formations. That would then reduce the amount of time spent undertaking sometimes complex manoeuvres in the vicinity of the enemy. When he realized he was approaching an ambush laid by Jugurtha, Metellus halted his army and altered the marching formation to one more similar to his intended battle line. He did not then have to stop again and redeploy in order to face Jugurtha in the ensuing pitched battle.[53]

The correlation between line of march and line of battle is clearly visible in Polybius' description of his second type of marching formation. This system could easily have been adapted for use by a legion organized in cohorts to form the *triplex acies* marching organization used by Caesar. We can also see the close correlation in the marching formations above **(15)**. A common formation for insecure situations or when anticipating battle had cavalry at front and rear, legions in the centre and usually auxiliaries between the cavalry and legions at van and rear. This could easily be converted into a standard battle line with legions in the centre flanked by auxiliaries and with the cavalry on the wings, in just the way that Polybius' marching formation for insecure circumstances could be transformed. Cavalry which acted as flank guards to the line of march would have had an important role in protecting the infantry whilst they manoeuvred into battle line. The cavalry screen could then quickly redeploy to the wings once the infantry was in place. This would allow the infantry an opportunity to deploy in relative safety, and was how Arrian intended to protect his infantry whilst they moved from line of march to line of battle in proximity to the powerful cavalry of the Alani.

None of the treatises explain how a line of march should redeploy into line of battle, and nor do any of the histories provide sufficient details for the manoeuvres to be deduced. Historians do sometimes mention that a marching formation could, or did, wheel into battle line, or make a single turn to form the line, but we are given no further details. Although Arrian too fails to explain exactly how he intended to redeploy his army from march to battle line, sufficient details of both proposed formations and the position of the units within them survive for an attempt at reconstructing the process. Arrian's marching column could have wheeled in either direction since he had arranged the army so that it could face a danger from any quarter. However, the column probably wheeled to the right to deploy so that Legion XV would hold the right wing of the heavy infantry in battle line. The right wing was traditionally the position of honour in the battle line and Legion XV was more likely to have held it since it was present in full strength, whereas Legion XII was represented only by a vexillation.

Arrian intended to face the Alani under particular topographical circumstances, anchoring his battle line with the infantry wings holding two hills.

1 The cavalry formed a screen round the infantry to protect them during the redeployment. Since the cavalry were already stationed on all sides of the marching column this would have been straightforward.
2 The infantry wheeled to the right: Cohors I Italica in the van took up position on the right wing; Cohors I Apulorum in the rear took the left wing. The legions in the centre of the marching column held the centre of the battle line.

3 The provincial militia and artillery joined the auxiliary infantry on the elevated ground at the wings.
4 The foot archers withdrew from the line of march and redeployed in a line to the rear of the legions.
5 The legions formed the main battle line, doubling the width of the marching column to create a battle line eight ranks deep, and filling the terrain between the two hills on which the wings were anchored.
6 A detachment of one hundred archers from Cohors III Cyrenaica joined the left wing on the rising ground to provide additional firepower.
7 The *equites legionis* and *equites singulares* also withdrew to the rear of the line, along with picked infantry, to serve as Arrian's bodyguard and as reinforcements.
8 Once the infantry were fully deployed, the cavalry moved to their assigned positions on the wings behind and to the side of the auxiliary infantry, and the mounted archers to the rear of the main battle line **(21)**.

It is clear that Arrian's line of march was carefully thought out with his intended battle line in mind. This must have been an important consideration for any general when drawing up a line of march which might have to deploy into battle line. It is not surprising then that when Germanicus advanced against the Cherusci in AD 16 with the intention of sending his auxiliary infantry into battle first, they led the army so they could enter battle immediately without the need for elaborate redeployment **(15)**. When an army was expecting to enter battle and was marching in *agmen quadratum* or a variant of that formation, the position of units in the intended battle line would have had more influence on the organization of the line of march than any other factor.

Logistics and the march

The length of the marching column and speed with which it could cover ground depended partly on the size and composition of the army, particularly its baggage train and the amount of supplies it was carrying with it. A large and heavy baggage train with many wagons would slow the army considerably by the speed of the animals and the requirement to provide extensive amounts of fodder for them, and would also limit the kind of terrain through which the army could progress. The Roman army campaigning in northern Greece in 169 BC seems to have relied on pack animals to carry supplies and equipment rather than wagons, a necessity because of the mountainous nature of the terrain. Even the pack animals, probably mules, experienced major difficulties and suffered serious damage when forced to retreat down a steep ridge and a series of wooden walkways had to be laid for the elephants to get them down the steep slope.[54] Roman armies did use wagons for transportation in more suitable terrain, and some are illustrated on Trajan's Column, but commanders regularly attempted to cut down the amount of baggage being carried by the army in order to speed it up:

> In order to reduce the size of the baggage train, which was greatly impeding the march of his army, Gaius Marius had his soldiers fasten their kit and rations in

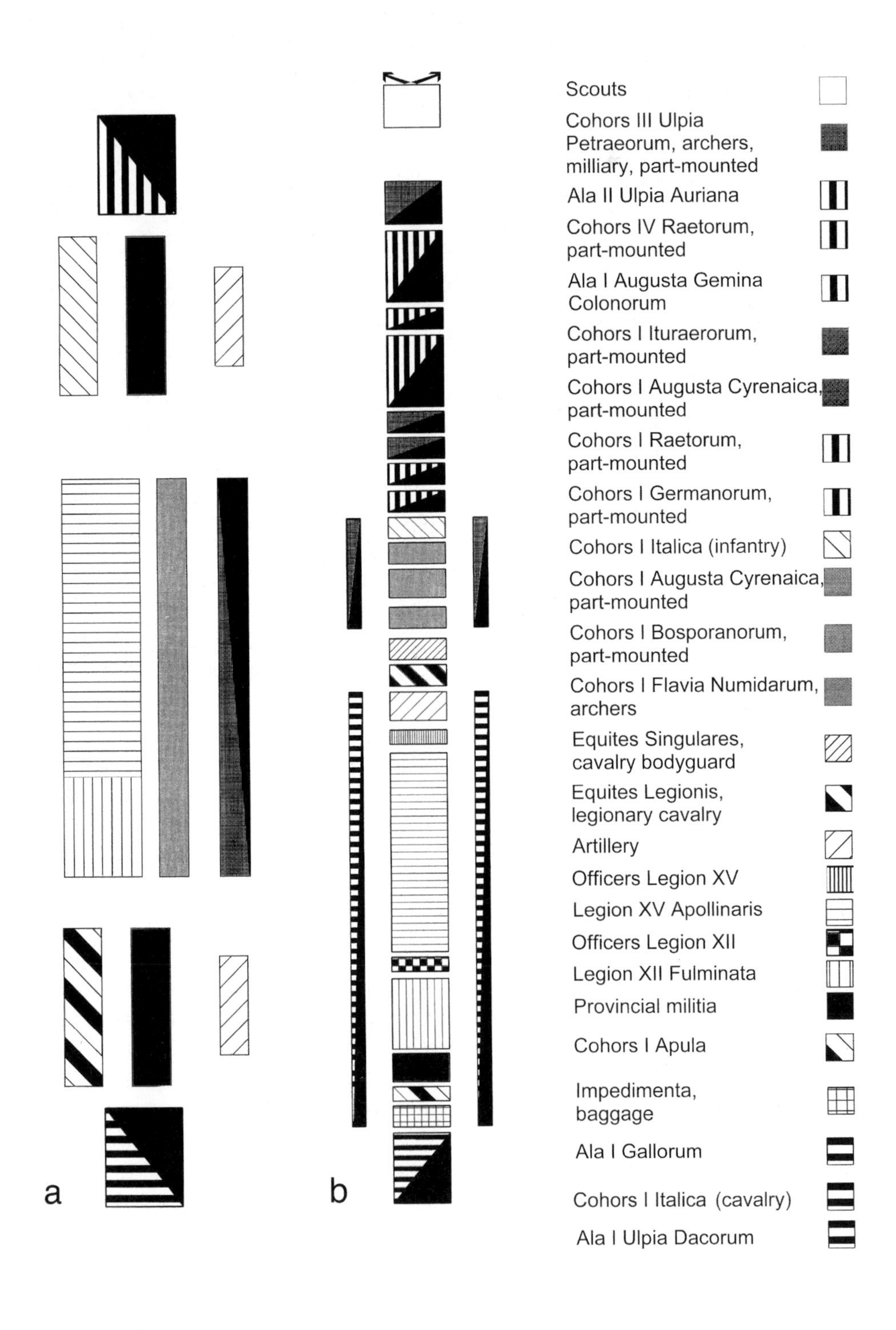

21 Arrian's battle order (a) and order of march (b) against the Alans

22 *Ox carts accompanying the army, Column of Marcus Aurelius.* Courtesy DAI

> bundles and hang them on forked poles to make the burden manageable, and resting easy. This is the origin of the expression, 'Marius' Mules'.
>
> Frontinus, *Stratagems* 4.1.7

Vegetius suggested that recruits should carry loads of up to 60 Roman pounds (nearly 20 kg) during training. This does not necessarily mean, though, that soldiers' packs normally weighed this amount. In their training, according to Vegetius, soldiers practised with shields and swords that exceeded the normal weight of equipment, and it is entirely possible that they did the same with their packs. Roman soldiers did carry with them a considerable amount of kit in addition to their armour and weapons. Josephus claims that each soldier carried a saw, basket, axe, pick, strap, billhook and chain, as well as three days' rations. Much of this equipment would have been necessary for entrenching camp, but since not all the soldiers were actually involved in the construction of the defences, with some of them providing protection for those labouring, it is unlikely that every soldier carried all this equipment. Estimates of the total weight of the soldier's load, including his arms and armour, range from 46-64.5 lb (21-29.4 kg), but the evidence is simply not available for any accurate assessment of what the soldier was expected to carry with him and how much it might have weighed.[55]

23 *Roman soldiers marching out, their kits slung over their shoulders on poles.* Courtesy JCN Coulston

Heavier items of equipment, including the leather tents used by the army, would have been carried on pack animals or by wagon, and the latter would have been necessary to transport artillery or a siege train, or supplies for more than a few days **(24)**. These animals would have slowed down the progress of the army, and their need for fodder may have required the army to camp earlier than otherwise, to ensure the animals were properly supplied by foraging parties. The presence of these animals would also have increased the numbers of servants with the army to look after them, further swelling its size and the numbers to be fed. It is not surprising then, to find generals like Marius attempting to lighten their armies either permanently or temporarily to speed up operations. Metellus had abandoned part of his army's baggage and loaded the pack animals with water to ensure his army would be able to cross fifty miles of desert to capture the town of Thala by surprise in the Jugurthine War. His son similarly took a very lightly loaded army, with supplies for only five days in an attempt to take the Spanish town of Langobrigae during the war against Sertorius. Cestius Gallus abandoned his pack animals and baggage to speed up his retreat at the start of the Jewish revolt in AD 66.[56]

Travelling lightly brought speed and independence, and in some of the cases mentioned above, surprise by which the objective might be captured, but it also brought the dangers of being detached from one's normal supply lines. Vegetius notes that in preparing for war, consideration should be given to supplies. Fodder, grain and other provisions should be requisitioned in advance and assembled at well-fortified places suitable for the waging of war. Papyrological evidence illustrates well the importance of local requisitions for the army in Egypt, and such sources could supply an army in peacetime or in war. For an army

24 *Mule carts, Column of Marcus Aurelius.* Courtesy DAI

campaigning outside the Roman Empire, requisitions could be made from allied tribes, though Onasander advises against remaining too long in allied territory because their crops would be quickly consumed. He recommends that an army should quickly enter enemy territory and make use of supplies found there. The allied kingdom of Cappadocia provided supplies for Lucullus in his campaigns against Mithridates in the 70s BC, whilst Tacitus mentions the impact of requisitions from Gaul for Germanicus' campaigns against the Germans in the early Empire, denuding the province of horses. In non-Roman territory, confiscations from locals, or purchases, might provide an important local source of additional supplies.[57]

Supplies gathered in these ways would probably have been stored at one or more supply bases, as Vegetius suggests, protected by a garrison, and stores from these shuttled to the campaigning army. After capturing the Numidian town of Vaga, Metellus collected corn and other materials there, making it a supply base and expecting that the many Italian merchants in the town would assist in the acquisition of supplies. He also installed a garrison to guard the base. After their defeat at Pharsalus, the Pompeians established a supply base at Utica in Africa with a view to continuing the resistance to Caesar in that area. Caesar meanwhile appointed Sallust, the historian, who was supporting him, to ensure his supplies for the coming campaign. Sallust took the island of Cercina after its Pompeian commander had fled and sent supplies of corn on to Caesar's camp on the

African mainland. Fortified supply bases can also be seen in the archaeological record: the fort of South Shields near Newcastle was converted in the early third century into a supply base with 22 granaries built in the fort. This was probably in preparation for the campaigns into Scotland of the emperor Septimius Severus, and South Shields seems to have been one of the principal supply bases for this expedition. Later it probably served as a base for supplying the Hadrian's Wall forts. The legionary fortress at Carpow at the mouth of the Tay may well have acted as a supply base as well as, perhaps, the headquarters for Severus' campaigns. Both South Shields and Carpow could have been easily supplied by sea, and supplies then transported overland to the campaigning army.[58]

As we have seen, armies might detach themselves temporarily from their supply lines to provide tactical independence and the possibility of moving far more swiftly to ensure an element of surprise, but if disruptions occurred to an army's normal supply, it could quickly find itself in difficulties. It was a valid tactic to target the supply lines of one's enemy, and this is exactly what Sertorius did when conducting successful guerrilla warfare against the armies of Metellus and Pompey in Spain. It is surprising then that the theoretical writers neither suggest this tactic, nor warn against it, but as already noted they provide very little advice indeed on supplying armies. They do, however, deal with the dangers involved in foraging.

Onasander stresses the importance of maintaining discipline whilst troops are out foraging or plundering, since such parties are vulnerable to attack. He advises that the foragers should be escorted by both cavalry and infantry. Their function was solely to guard those soldiers involved in the collection of food and fodder. This seems a sensible precaution, but even so it is perhaps surprising the number of times Roman armies were caught when foraging and got into difficulties. The Britons had attacked one of Caesar's legions whilst it was foraging in 55 BC, setting an ambush at the only patch of grain in the area that the Romans had not yet harvested. Since Caesar had no cavalry on this expedition, he was unable to provide proper protection for his foragers. Sertorius ambushed one of Pompey's legions which had gone out to forage, along with some camp servants. By preventing access to one of only two areas from which forage could be gathered, Sertorius was able to set an ambush to catch Pompey's troops when they had completed their task and were loaded down with forage. The guards whom Pompey had sent out with his foragers had lost concentration and were slipping off to do their own foraging when the ambush was sprung; Pompey suffered a serious setback when his reinforcements were also defeated. Frontinus cites Livy as reporting that Pompey lost 10,000 men and all his pack animals, a loss that would have added to his logistical problems. Ensuring the discipline of foraging parties then was just as important as maintaining discipline on the march. Appian contrasts the behaviour of the foragers of the two generals Manilius and Scipio Aemilianus during the siege of Carthage in the mid second century BC. Manilius' foragers were in the charge of the military tribunes and were not sufficiently alert, opening themselves to attack by African cavalry. But Scipio ensured the foraging area was fully scouted by both cavalry and infantry, and then guarded by them, and that those foraging always remained within the protective circle. Scipio himself rode round the circle ensuring that the foragers remained within it, and punishing those who did not.[59]

25 *A legionary foraging; his bronze sickle no longer survives; Trajan's Column.* Courtesy JCN Coulston

The requirement of supplying an army and procuring forage and fodder for the baggage animals and cavalry horses could then have had a considerable impact on the composition and speed of the army's marching column. It could also influence the direction of the army's march and even dictate strategy. An army might be diverted from its objective by the need to ensure supplies, or forced to take a particular route. On the other hand, a general might deliberately advance through adverse terrain in order to take the enemy by surprise, as Metellus did in Africa to take Vaga. Corbulo too marched his army through hot and waterless mountain regions of Armenia to terrorize the enemy in distant parts of the country. Alternatively, an army with inadequate supply lines, or short of supplies, might be forced to divert and attack an enemy stronghold in order to obtain them. Caesar captured the Thessalian city of Gomphi to alleviate his army's serious problems with supply, and his decision to force battle with Pompey in the area of Pharsalus was dictated partly by the plentiful supplies of forage and corn in the area.[60]

Conclusion

A line of march was far more than simply a means of getting an army from one place to another. The marching camp provided a secure encampment for the army each night. During the day, a well-organized column advancing through properly reconnoitred terrain could have provided a similar security for the army, though this no doubt encouraged the mental dullness that Vegetius and Onasander warn against. Perhaps it is not surprising to find that on the long journey with his army to his Parthian campaign, Trajan varied the order of his army's march, and had false reports circulated of enemy movements to ensure his army remained on the alert.[61]

The organization of the marching column would depend on the tactical situation. Long, narrow columns were generally avoided if there was any indication of enemy presence. An aggressively deployed army marching in *triplex acies* formation could have been an impressive and intimidating sight. It also prepared the army for facing the enemy, in ambush or pitched battle. In the latter case, provided the general had planned both his marching formation and battle arrangements carefully, the dangers involved in redeployment could be reduced. There are many examples, however, where such plans were not made, scouts were not deployed, and the army was wiped out by ambush. The Roman commander at Trasimene, Flaminius, had previously enjoyed a successful military career and a triumph, yet in 217 BC he failed to take the basic precautions which might have saved his army. Caesar too was ambushed several times in Gaul, though his prepared army suffered no serious harm.

Although Roman armies could move at speed if necessary, covering up to 25 miles in a single day, like most pre-mechanized armies they generally moved slowly. The comparatively short distances covered in a day's march would have made the task of logistical support easier, but campaigns longer. Speed then could be used as a weapon to take the enemy by surprise, as Cato did in Spain.

3 At rest: campaign camps

Introduction

> Your ancestors considered a fortified camp to be a haven against all the misfortunes of war: a place from which to go out to fight, and in which they could find shelter after being tossed by the storms of battle. Thus, when they had fortified their camp, they strengthened it with a strong guard, because anyone forced from his camp, even if he had won the battle, would be judged to have been defeated. A camp is a shelter for the conqueror, a refuge for the conquered. How many times has an army, after meeting less than favourable fortune in battle, been driven within their ramparts and then in their own time, sometimes after only a moment, sallied out against the victorious enemy and repelled him. This place is the soldier's second homeland, with ramparts for city walls, and his tent is his hearth and home.
>
> Speech of Aemilius Paullus before Pydna, 168 BC (Livy 44.39)

According to Polybius, the Roman army on campaign surveyed and entrenched a camp each night, and other authors claim that the Romans never fought a pitched battle without first fortifying an encampment. Such encampments were temporary, occupied normally for a few days at most and often for just one night. Despite its temporary existence, the marching camp, or temporary camp, is one of the most potent symbols of the discipline and organization of the Roman army. It is also, in archaeological terms, one of the clearest indicators of Roman military activity. The remains of these camps can provide information about campaigns for which little or no literary evidence survives, and can give some indication of the size of the Roman army involved in a particular campaign.[62]

The marching camp had several functions. In pitched battle, as Livy explains, it provided 'a shelter for the conqueror, a refuge for the conquered'. After Cannae, some 17,000 Roman survivors are reported to have reached the safety of the two Roman camps. The Carthaginians did not attempt to take the camps by storm, concentrating instead on mopping up the Romans who had escaped to the undefended village of Cannae. During the night, the survivors in the two camps united, and some of them later escaped to the safety of the nearby town of Canusium.[63] The camp could allow survivors of a defeated army the opportunity to regroup and, as Livy suggests, possibly to renew the battle against the enemy. In pitched battle, the vast majority of casualties occurred to the defeated army once its battle line had turned and was in flight. The presence of a fortified refuge near the

26 *Legionaries fortifying camp using turves, Trajan's Column.* Courtesy JCN Coulston

battlefield could allow the fleeing soldiers to reach shelter comparatively swiftly. This could reduce the extent of the losses, even if the defeated side did not then sally and reverse the initial result. The positioning of marching camps in relation to intended battlefields is therefore of considerable importance, and this was acknowledged by the Romans. During the Civil War campaign in Spain, the Pompeians under Afranius offered battle, and although Caesar deployed his troops in response, he did not wish to commit himself to battle at that point. He was aiming to force the Pompeians to surrender rather than risking his men in pitched battle, a sensible strategy given that he already had the upper hand. In any case, Caesar tells us, the nature of the battlefield was such that a decisive victory was unlikely:

> The confined space of the battlefield would not help in winning a decisive victory, even if the enemy were routed. For the two camps were not more than two miles apart. The two armies occupied two thirds of this space; the remaining third was empty, left for charging and attacking. If battle commenced, the proximity of the camps would allow the beaten side to flee and find refuge quickly.
>
> Caesar, *Civil War* 1.82

But at Pharsalus, where both sides were prepared to fight a decisive battle, one account claims that Caesar gave the extraordinary order to his men to destroy their camp fortifications before they deployed. Appian reports this action, and points out that for Caesar's men the battle was a fight to the finish: there would be no retreat from battle and so they would fight with greater determination.[64]

An army that was properly entrenched at night would be less vulnerable to surprise attacks by the enemy. Whilst the fortifications of Roman marching camps were not designed to be defended for very long (they were in most cases comparatively small), they might give the defenders time to react to an attack and to deploy outside the camp. The effects of an attack on an unprotected army can be clearly seen when Dolabella made a dawn attack on the unfortified encampment of Tacfarinas in Africa. The Numidians were panicked by the sudden noisy arrival of the Romans, were unable to react in time, and were wiped out. Surprise attacks like this on Roman encampments are extremely rare; in the campaign against the Veneti, the Gauls under Viridovix attacked the camp of Caesar's legate Sabinus, but his troops sortied to meet the attack. The camp was situated at the top of a long, gentle slope, and since the Gauls had charged up it, they were too exhausted to fight, and fled without making a stand. This episode provides an excellent example of the use of the marching camp. The Gauls attacked the camp because they thought the Romans were too scared to come out and fight, but the camp's position and the timely sortie by the Romans led to a decisive victory.[65]

In the speech of Aemilius Paullus, Livy alludes to the role of the marching camp as the soldier's home for the duration of the campaign. The camp could provide the soldiers with a sense of security, with familiar surroundings wherever they were campaigning in the empire. The camp itself and its environment could also help to foster unit cohesion and a sense of belonging, to *contubernium*, century and beyond, to the legion and the army. Though Josephus and Vegetius compared the construction of a marching camp to a town mushrooming up, the camp actually served to distance the soldier from the civilian life of the town and stress his identity as part of the military community of the camp. Units that gave way in battle or showed cowardice might be punished by being ordered to camp outside the security of the camp fortifications, excluded from the military community they had let down until they had redeemed themselves.[66]

In an encampment with a standard layout and organization in which each soldier knew his allotted space, reaction time in the event of an alarm would be cut down. The security provided by the camp and its fortifications might enable soldiers to pass the night feeling relatively safe, even in hostile territory. At the very least, it would help to prevent wild animals from wandering into the camp and causing an alarm. The camp could also serve as a deterrent against desertion from the campaigning army. This may be hinted at by Pseudo-Hyginus when he claims that in secure territory physical defences were not always used, but that a line of pickets was deployed instead 'for the sake of discipline'.[67]

This 'home from home' of the Roman soldier could create a small trail of Roman occupation across the countryside. Although many Roman writers claim that the defences were slighted when the army left camp, the archaeological evidence shows clearly that in practice this was not always the case. This is not because armies expected to re-occupy the camps, but more likely because their destruction was simply considered unnecessary. The remains of these camps could provide a reminder, in both conquered provinces and outside the empire, of the discipline and presence of the Roman army. Despite their temporary occupation, marching camps could act as permanent symbols of the power of Rome.

Polybius and Josephus describe the campaign camp in their digressions on the Roman army, and both express admiration of the Romans for their thoroughness and efficiency at undertaking such a construction task on a regular basis. Josephus' description is sketchy, but Polybius included information about the internal organization of the encampment, precise measurements, and the allocations of space to particular sections of the army. Polybius may well have got his information on castrametation, the art of surveying a camp, from a handbook on the subject, the Republican equivalent of Pseudo-Hyginus' manual *de metatione castrorum*. The latter explains the laying out of a camp in considerably greater detail than Polybius and, unlike the earlier author, includes information about the location and defences of camps. In addition to this very useful information, we also have the archaeological remains of marching camps. Dozens of these have been located in Britain, principally through the use of aerial photography, but their survival is limited mainly to upland areas which have not been seriously disturbed by farming in the intervening period. A few marching camps have been identified in North Eastern France and on the Danube, but the vast majority of the archaeological evidence comes from Britain, primarily because that is where most exploration has taken place. The Roman siege camps at Numantia in Spain and Masada in Israel can also help our understanding of the marching camps since the encampments, built in stone for longer occupancy, may nonetheless reflect the organization of the temporary structures.

The origins of marching camps

The marching camp with its standardized internal organization is considered by many modern writers to be distinctly Roman. The camp is seen as a reflection of discipline, one of the principal reasons for Roman military superiority. Indeed, Corbulo, one of the foremost generals of the early Roman Empire, once declared that the pick was the best weapon to use against the enemy. Despite the distinctly Roman image of the marching camp, however, there is some confusion amongst writers of the Roman period about its origins. Frontinus claims that Pyrrhus of Epirus was the first general to concentrate his entire army within the same fortifications in the field, and that after capturing his camp at Malventum in 275 BC, the Romans adopted his method of entrenchment. Livy agrees that Pyrrhus was the first to teach the art of castrametation. On the other hand, Plutarch describes how, on spying out the Roman camp, Pyrrhus was surprised at the discipline of his enemy and the arrangement of their camp; he realized that he was not fighting barbarians.[68]

The use of a defended encampment was certainly not something that was unique to the Romans. Before the battle at Plataea in 479 BC, the Persians fortified a camp with a wooden palisade to provide a refuge should they be defeated by the Greeks. The camp was constructed for the same purpose as Roman camps before pitched battle. After their defeat, some of the Persians reached the fortified camp and successfully defended it against a Spartan attack; it was not until the Athenians arrived that the camp was captured (Herodotus states that the Spartans were useless at attacking fortifications). We hear

27 *Egyptian encampment under attack at the Battle of Qadesh, thirteenth century BC*

nothing about the internal arrangement of the camp, but the Romans were not alone in employing a sophisticated method of organising their camps. As early as the thirteenth century BC, the Egyptian army was camping in the field in an organized fortification. The relief of the Battle of Qadesh from the temple complex at Abu Simbel in Egypt shows the rectangular camp protected by a row of shields **(27)**. Two main roads bisect the camp, and where they intersect is a building, possibly a headquarters or shrine, or possibly both. Fortified camps, usually in the context of siege warfare, are frequently illustrated in Assyrian reliefs of the ninth and eighth centuries BC. Again, the camps have two principal roads meeting at a headquarters tent, or the king's tent **(28)**.[69]

The Greek historian Xenophon says that it was the standard practice of the Persians to fortify their camp with a ditch, implying that this was done on a nightly basis. He describes in detail the method of camping employed by the Persian king Cyrus: the king's tent was in the centre of the camp with various troops arranged around it in an organized fashion:

> Everything else was organized in such a way that each man knew his own place in the camp, both its size and location .... He (Cyrus) positioned himself in the middle of the camp, because this was the most secure place; then came his most trusted followers ... and beyond them in a circle were the cavalry and charioteers. Finally, the hoplites and those armed with the large shields surrounded the whole encampment like a wall.
>
> Xenophon, *Cyropaedia* 8.5

Roman writers frequently mention that all the soldiers knew their place in the camp and, again like the Roman method of encampment, each soldier in Cyrus' camp had a designated space of particular size within the camp. The tent of Cyrus himself was

28 *Assyrian encampment, relief from Nineveh, ninth century BC.* Courtesy The British Museum

surrounded by his most trusted followers, like the Praetorian Guard around the Emperor's tent in a Roman camp. These observations though may well be more relevant to Xenophon's own day (the fourth century BC) and to his own military theories than to the Persia of Cyrus the Great two centuries previously. But the details of the careful arrangement of the camp described by Xenophon suggest that some logical method of camp organization was in use or available in theoretical form by the fourth century BC, if not before. Xenophon's contemporary Aeneas Tacticus wrote a treatise on castrametation which is now lost, but it included advice on the posting of guards and patrols. It probably also discussed fortifications and internal arrangements. Whether or not Cyrus actually used the castrametation techniques Xenophon claimed he did, it is nonetheless interesting that the historian attributes them to a Persian king. There certainly seems to be a very long tradition of organized encampments in the civilizations of the Eastern Mediterranean, and this may be a direct forerunner of the Roman system.[70]

Livy mentions Roman fortified camps as early as 479 BC, but he may be thinking of the standard military procedures of a much later period and assuming that they had always existed. It is impossible to know when the Romans first started using the design of camp that Polybius describes, though an organized method of encampment was certainly in use by the Second Punic War at the latest (218-202 BC). The stories about Pyrrhus and the Roman camp indicate an earlier date for their use, in the early third century BC, but it is not until the second century BC that we have good evidence for their design. This is in the work of the Greek author Polybius, and in archaeological remains in Spain. Polybius, who expresses great admiration for Roman military institutions in general, is particularly impressed by the organization and defences of the Roman camp. He contrasts the Greek and Roman methods of encampment, criticizing the Greek and praising the Roman:

> The Greeks in encamping think that the greatest security lies in fortifying the natural strength of the position, firstly because they shun the labour of entrenching, and secondly because they think that man-made defences are not

> as strong as those provided by the natural features of the site. And so as regards the plan of the camp as a whole, they are forced to use all kinds of shapes to suit the nature of the terrain, and at times to move part of the army to unsuitable locations. The result is that everyone is uncertain of his place in the camp and the location of his unit. The Romans on the other hand prefer to submit to the fatigue of entrenching and other defensive works for the sake of having a consistent and uniform plan for the camp that is familiar to everyone.
>
> Polybius 6.42

As already suggested, Polybius probably got his information for Roman camping techniques from a textbook. This book would only have told him how its author thought things were supposed to be done, not how they were actually done. In practice, the organization may not have been as consistent and uniform as Polybius boasts. At Renieblas in northern Spain, near to the Celtiberian town of Numantia and the modern city of Burgos, are the remains of a series of five Roman campaign camps **(29)**. The defences of the earliest camps (I-III), dating to the mid second century BC, follow closely the contours of the hill and so have irregular outlines. However, the later camps (IV-V), which may date to the campaigns against Sertorius in the late 80s BC are rectangular, with Camp V surveyed along the steep slope of the hill. The defences of the early camps here may follow the contours of the hill because they were constructed under less secure conditions than the later camps, but the shapes of these early camps meant that the units could not all camp in the organized way that Polybius describes in such detail. In Camp III, lack of space may have meant the allies were not located in their usual area, but in an annexe to the main camp. The early camps at Renieblas may be linked quite closely typologically with the Greek defended hilltops that Polybius criticized. Indeed, the series of camps here may illustrate the development of Roman camps from the defended hilltop to the standard rectangular form more usually associated with the Romans.

The nightly construction of a fortified and organized encampment was not something unique to Rome, nor was the building of such a camp before fighting a pitched battle. Such camps had been used in Egypt, Assyria and Persia well before the Roman period, and the Greeks of the fourth century BC produced textbooks on the art of castrametation. Although the absence of reliable data demands caution in trying to establish the origins of Roman marching camps, it may well be that Frontinus and Livy were right when they stated that the Romans adopted their form of camping from Pyrrhus.

Location of camps

Historians and archaeologists have not previously considered in much detail the location of campaign camps and the factors that governed their locations. For the Romans, however, this was clearly a matter of great concern. The writers of ancient military theory provide extensive advice on the subject, and Roman historians frequently comment on the location of camps. The ability to choose a good campsite was considered by the Romans to be one of the qualities of a good general. Caesar, Vespasian, Agricola and even Hadrian, though he never actually went on a real campaign, were all endowed with this ability, if we

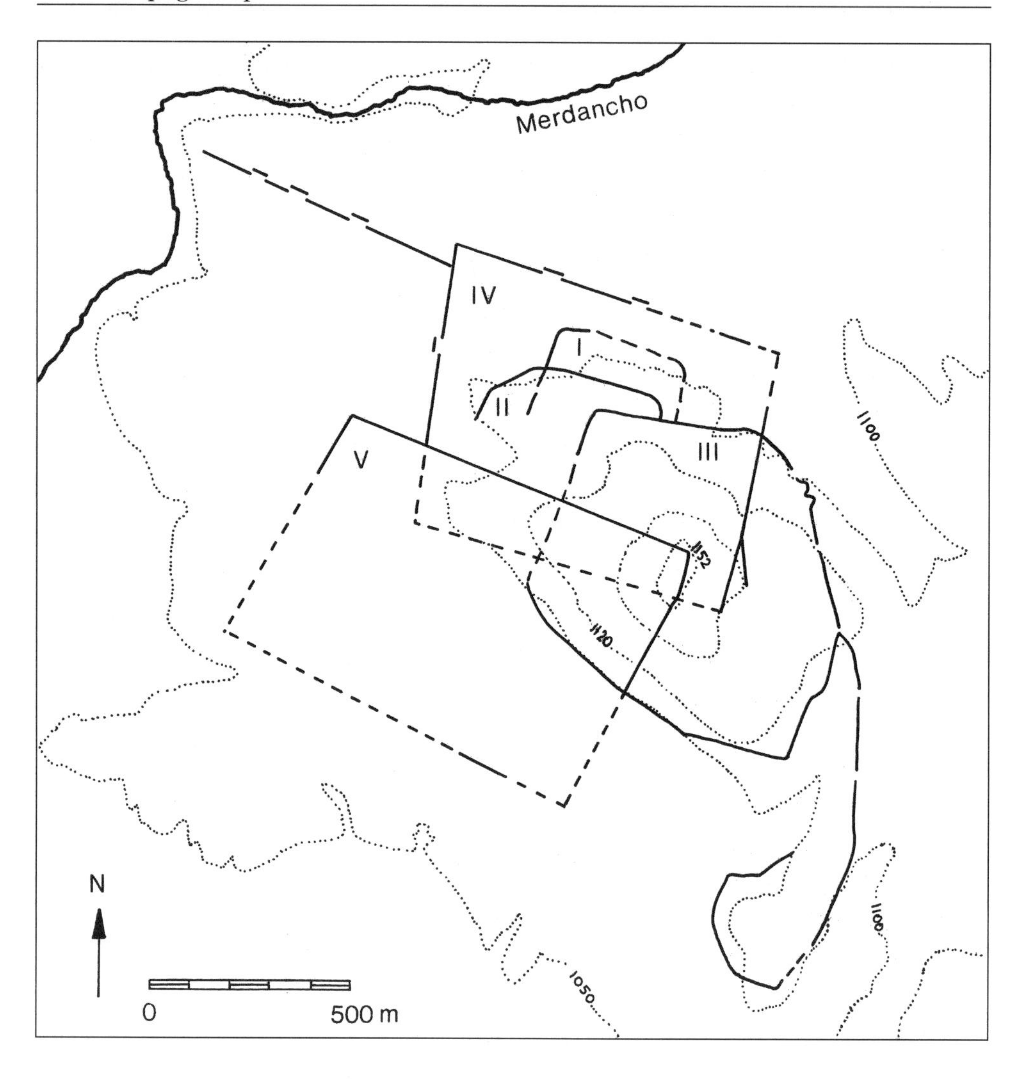

29 The Roman encampments at Renieblas near Numantia, Spain (after Schulten)

are to believe the various historians. Moreover, we are fortunate in having such a wealth of archaeological evidence for temporary camps. By comparing these remains with the theory of the textbooks, we can compare the theory with the field practices of the Roman army.[71]

The treatise writers list a number of factors that had to be taken into account when surveying a suitable campsite. Temporary camps should be near to supplies of wood and grain, and to a water supply, but should not be on marshy ground or on land liable to flood. The site should be healthy, particularly if it is to be occupied for more than a few days, and Vegetius advises against remaining too long in the same camp because the air and water will become polluted. The camp should not be overlooked by higher ground; it should not be near to forests, gullies or valleys that might assist the enemy in a sudden attack. The *porta praetoria* should always face towards the enemy, or down any slope, or in the direction of the intended march.[72] There are other factors too, which must have been considered when surveyors chose the site for a camp (Vegetius suggests the job was done by surveyors, not generals). The proximity and nature of the enemy must have been taken into account, and the location of a potential battlefield, if pitched battle was likely. The size of the army was also of significance: surveyors had to ensure that the chosen site would be large enough to fit the whole army. This task could have been very difficult and may have affected the distance marched in a particular day in order to reach a suitable camp site. Pseudo-Hyginus labels camps that are located on poor sites *novercae*, literally 'stepmother', though perhaps 'mother-in-law' would convey the sense better to a modern audience. These positions, he says, should be avoided if at all possible.

It is not surprising that surveyors or generals were not always able to fulfil these requirements. Appian neatly illustrates a lot of the advice, and failure to take it, in his comparison of the camp of Brutus and Cassius with that of Antony in the Philippi campaign of 42 BC:

> Immediately the inferiority of one camp and the superiority of the other became apparent: Brutus and Cassius were camped on the hills, Antony on the plain; the former obtained firewood from the hills, the latter from the marsh; the former drew their water from a river, the latter from newly dug wells; the former obtained their supplies from Thasos only a few stades away, but Antony was 350 stades away from Amphipolis. It seems, however, that Antony was forced to do these things, for there was no other hill and the rest of the plain, being low-lying, was sometimes completely flooded by the river.
>
> Appian, *Civil War* 4 107

Antony, it seems, had a veritable 'mother-in-law' of a campsite, whilst his enemies had seized the best place in the area for their camp.

Although the treatises recommend camping near to supplies of wood, grain and water, it is the latter that seems to have been most important. This is not surprising, since the health of the army including horses and baggage animals depended on adequate supplies of water. Commanders seem to have made every effort to camp near a water supply and conversely, to force the enemy to camp where there was no water, or an inadequate supply. We have seen how in the civil war campaign in Africa, Labienus harassed Caesar's army whilst on the march to force him to camp where there was no water. Later in the same campaign, Caesar was obliged to camp further from the enemy than he wanted, in order to be near an adequate water supply. So important was access to water that when necessary, additional ramparts might be constructed to secure a line between a camp and the nearest water supply. During the civil war campaigns in Spain, Petreius and Afranius had such a rampart, or *bracchium*, constructed from their camp to the river because those getting the water were being harried by Caesar's cavalry. The Roman army occupying Camp IV at Renieblas also built a *bracchium* to ensure communications between the camp and the river Medencho to the north-west. The fortified arm is an extension of the rampart and would have prevented sudden or casual attacks on those getting water. Marius, however, is supposed to have deliberately chosen a site for his camp some distance from a water supply. Plutarch reports that the Roman camp was in a strong position, but the nearest river ran near to the encampment of the Cimbri and Teutones. The attempts by the Romans to get water precipitated an engagement with the enemy. Plutarch, though, may be interpreting a situation that had arisen by accident as one engineered by the shrewd Marius.[73]

Clearly access to an adequate water supply was of great importance in choosing a campsite, perhaps of primary importance, overriding other factors such as proximity to the enemy. However, other topographical factors did have to be considered. As well as being vitally important, water could also be a source of potential danger. The treatises warn against camping on land liable to flood, and the dangers of ignoring this advice are well illustrated during Caesar's campaign in Spain against Petreius and Afranius. He was forced to camp his army between two rivers and when they flooded, the army was cut off for several days without adequate supplies of food. During the Batavian revolt of AD 70, the Romans appear to have had no choice but to camp on flat ground, and when the Rhine flooded, the camp was washed away.[74]

Other potentially dangerous places to locate a camp were near to forests and gullies by which the enemy could approach unseen. It was not always possible to avoid such features though, and tree clearance, if undertaken at all, would be a very arduous task. At one camp, Caesar was taken by surprise because he had camped on the edge of a forest and the Gauls were able to advance under cover. Caesar claims that his men reacted quickly, armed themselves and drove the Gauls back into the forest, and it is likely that the soldiers kept their weapons close by when engaged in this work. Soldiers illustrated entrenching a camp on Trajan's Column are shown as having piled their weapons and armour nearby so they would be easily accessible in case of a sudden emergency **(34)**.[75]

The archaeological evidence illustrates well that on the whole the advice on the siting of camps in the theoretical works is sound. The vast majority of temporary camps in Britain are situated within very easy reach of a river or other water supply, and many are on rising ground beside a river or other water supply. This usually ensured an easily defendable site and some visibility as well as the water supply; few camps are far from water. Other factors also influenced the location of marching camps: they could be sited to command river crossings and fords, valleys and mountain passes. The tendency to construct camps in Wales on plateaux means they are often further than usual from water, but such was the compromise to ensure good visibility, often on all sides, such as Y Pigwn near Brecon **(30)**. The camps at Troutbeck in Cumbria commanded the valley of the River Eden, whilst North Tawton in Devon covered a crossing of the River Taw. The position of Malham in North Yorkshire is not a good defensive one, being close to gullies on two of its sides, but has good visibility instead, and Swine Hill in Northumberland has good views but a gully on one side.[76] One disadvantage, particularly with large camps, was that it may not have been possible to have an uninterrupted view from one side of the camp to another. This was very likely if the camp, like many in Britain, enclosed knolls of slightly higher ground, such as Featherwood East and Silloans in Northumberland. Under such circumstances, the army's trumpets and horns would have to be relied on to pass signals. This may be one of the reasons that Pseudo-Hyginus recommends the specific

30 *Two Roman camps at Y Pigwn, Brecon. The internal* claviculae *can clearly be seen protecting the gateways.* Courtesy Cambridge Committee for Aerial Photography

proportions for marching camps that he does, because then as he points out, both instruments can be heard.

The surveyors aimed to make the best use of the terrain when laying out camps. Their location and defences were probably often chosen and surveyed with this in mind. Some camps appear to ignore the advice of the treatises about camping too near to gullies, but in these instances the location actually enhances the defensive capabilities of the camp. Greenlee Lough in Northumberland is one example: the west rampart is protected by a steep sided gully. The fortifications of some camps were extended to maximize the defensive potential of the terrain, in the way that Greek encampments and the earlier camps at Renieblas did. Consequently, camps are not always regular shapes; Raedykes has an irregular extension in one corner to enclose ground that would strengthen the defences. Not all irregularities, though, can be explained away like this, and some irregular plans, like Pennymuir, may be the result of surveying errors instead. Most camps seem to have been sited to enclose any high ground or knolls within the immediate area, or if it was not convenient, an annexe might be added on for the same purpose. Little Kerse has such an annexe, and the annexe of Camp III at Renieblas protected a steep sided gully, which might otherwise have enabled an enemy to approach the camp unseen. It is rare to find a camp that is overlooked by higher ground: Ythan Wells in Scotland is an exception, and Livy reports a camp situated beneath a hill, but the higher ground had been captured and presumably secured with a small garrison to ensure the safety of the camp below.[77]

Marshy terrain, as advised by the treatise writers, is generally avoided, though occasionally a camp will enclose some wet ground. Arosfa Careg, Esgairperfedd and Chew Green III all include some marshy or wet ground. Interestingly, some practice marching camps were also built on wet terrain, possibly to give the soldiers experience of building camps in difficult circumstances: one of the practice camps at Gelligaer Common in South Wales encloses marshy ground. The inclusion of poor ground within camp defences must have been unavoidable at times, particularly if the army being encamped was a large one. Obviously, the larger the army the harder it must have been to find a large enough area of suitable land that fulfilled the various topographical and tactical criteria that the treatises list. Areas of 'dead space' may have been included within the perimeter of the camp under such circumstances, and the overall size of the camp increased accordingly. Compromises must have been made, and so the advice of the treatises was not always followed to the letter.

Camp defences

Whilst the location of a camp was usually influenced by topographical considerations, the type and size of its defences would depend on the nature and location of the enemy. As with the order of march, standard arrangements might be altered in the vicinity of the enemy or when travelling through hostile terrain. Under some circumstances, physical defences for a camp might have been considered unnecessary, though such practices are frequently criticised by ancient writers as being too risky. Sallust's general criticisms of the Roman army in Africa before the arrival of Metellus include the accusation that the camps were not fortified. With Metellus in command, however, camps were entrenched every night and each one was fortified 'with a rampart and trench as if the enemy were close at hand.' Antony, a very experienced general, is criticised by Appian for camping near to Lepidus (at that time his enemy) without palisade and ditch 'as though he were camping alongside a friend'. What Appian does not mention though is that Antony may have done this deliberately to facilitate the fraternization that was happening between his soldiers and those of Lepidus. Such actions could then make the soldiers less willing to participate in the fratricide of civil war, and Caesar had used this tactic too. Cerialis had no such intentions, however, when he camped for several days without rampart and ditch during the Batavian revolt. He only dug defences when Civilis and the Batavians neared, and Tacitus describes this lapse as 'imprudent'.[78]

Clearly then, commanders did not always fortify their camps. The historians might criticise commanders for camping without defences, but they did acknowledge that practices might vary according to the military situation. The writers of the military treatises recognized this fully. Both Pseudo-Hyginus and Vegetius indicate that camp defences could vary according to the potential danger as well as the type of soil being dug. Pseudo-Hyginus, as we have seen, goes as far as to suggest that in secure places a single trench or row of armed guards would be sufficient defence. Josephus confirms the idea that the strength of the defences could vary, stating that the camp was fortified with a trench *if necessary*.[79]

This could explain the absence of marching camps in some areas where Roman armies are known to have campaigned. Virtually no marching camps have been found in Britain south of Nottinghamshire. Some archaeologists have argued that this as the result of destruction through agriculture and building, and soil types not conducive to producing crop marks. However, the comparatively large numbers of camps elsewhere, particularly in areas of great strategic importance such as Shropshire, the invasion route into central Wales, might suggest an alternative explanation. Those areas of southern Britain without marching camps were over-run by the Romans within a very short time after the invasion, and the presence of pro-Roman or client kings in this area would have facilitated speedy and easy conquest. It is very possible that, like Cerialis and Antony, the legates campaigning in these areas considered fortifications unnecessary, or were content with very slight defences. One of Germanicus' campaign camps in Germany was defended by earthworks to front and rear, but only had palisades along the sides. Such fortifications would leave little or no trace archaeologically.

The principal components of a camp's fortifications were the rampart and ditch, the palisade, and additional defences to protect the vulnerable gateways. Few ancient writers are terribly interested in these kinds of details, and even Caesar does not usually bother to note such comparatively trivial things. He only twice mentions the dimensions of camp fortifications, and on both occasions an attack was expected and the defences were particularly strong. Caesar only includes this information because the circumstances and therefore the defences were exceptional. Although there is plenty of archaeological evidence for camps and their defences, it is impossible to know the circumstances under which they were constructed and as already indicated the strength of the fortifications could vary considerably.[80]

According to Pseudo-Hyginus, perhaps the most reliable source on marching camps, the ditch should be 5 Roman feet wide (c.1.5m) and 3 feet deep (c.0.9m), and it should be either V-shaped or Punic **(31)**. The Punic ditch, so-called because of the notoriety of Carthaginian or Punic perfidy, had asymmetric sides and could trick the enemy. The design could cause an attacker difficulty in getting out of the ditch or even trap him there, an easy target for the defenders. Punic ditches are, however, extremely rare. The only good examples are at the fort of Hod Hill in Dorset, and no marching camps have been found with ditches of this type. Their asymmetric sides probably made Punic ditches harder to cut, and perhaps the ordinary V-shaped ditch was considered sufficient labour for a camp that would be occupied for such a short time. Vegetius does not specify the type of ditch, but recommends different proportions for different situations: an ordinary encampment,

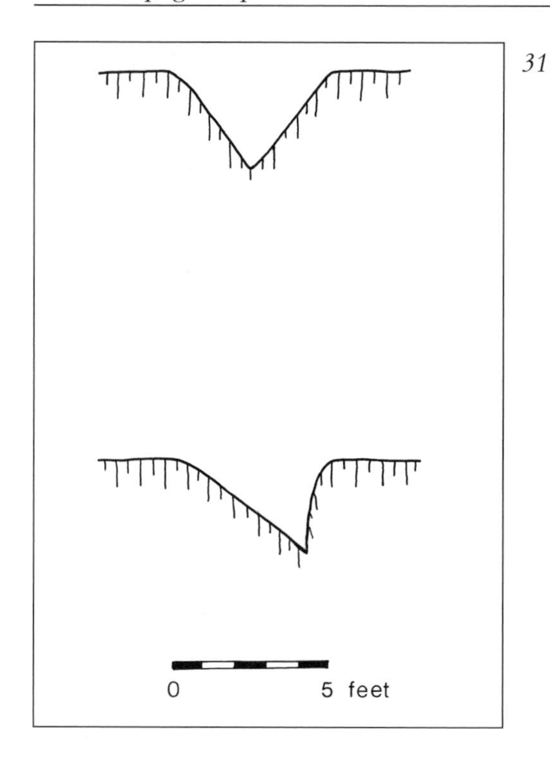

31 Ditch types: V-shaped ditch and, below, Punic ditch

a camp likely to be attacked, a camp for long-term occupation, and for one built on friable soil.[81] The last consideration is an important one, and the archaeological evidence illustrates well that the size of the ditch could depend on the soil type and geological factors. The ditches of several camps, including Durno, Malham and Rey Cross, are much shallower in places, or even interrupted, because of bedrock near the surface. At Gogar near Edinburgh, the ditch is fairly small on the clay soils of the western side of the camp, but wider and deeper on the more friable sandy northeastern side. Overall though, the size of the ditches of most marching camps in Britain corresponds well with the advice given by Pseudo-Hyginus. Some defences are considerably stronger, such as Raedykes in Scotland, and Bernhadsthal and Kollnbrunn near Carnuntum on the Danube, all of which have ditches up to six and a half feet (2m) deep and over ten feet (3m) wide. The most logical explanation for defences of this size is that the camps were at greater risk of attack, but this cannot be known for certain; they may have been built for longer occupation which would require more extensive defences.

The rampart was constructed from rubble or turves, most likely the fill from the ditch. As explained above, a rampart was not always considered necessary. Although Pseudo-Hyginus states that the rampart should be six Roman feet high and eight feet wide (c.2m x 2.6m), it is only necessary, he says, in less secure places. Vegetius, whose ditch recommendations are much deeper than those of Pseudo-Hyginus, suggests that a rampart three Roman feet high (one metre) was sufficient even for a camp likely to be attacked. The rampart then was not under normal circumstances the most important component of the defences, although a camp in hostile terrain would probably have had a strong rampart with, as Pseudo-Hyginus says, good access for the soldiers and artillery platforms. Construction of a rampart would also have been difficult in rocky terrain or desert conditions which Pseudo-Hyginus acknowledges. A form of stockade is recommended instead. Roman armies campaigning in sandy areas may have carried empty bags so a rampart of sandbags could be constructed. Vegetius states that the Persians (whom, he claims, copied the Roman method of camping) did this, and there is no reason to suppose that the Roman armies did not do so also.[82]

The ramparts were topped by a palisade constructed of wooden stakes, *cervoli* or *valli*. If the ground was unsuitable for a rampart to be constructed, or if a rampart was considered unnecessary, these same stakes could be used to make the palisade that Pseudo-Hyginus considered was a satisfactory alternative. The stakes could form an effective barrier, as Livy explains:

> The Romans cut light stakes (*valli*), generally two-forked, with three or maybe four branches, so that a soldier can comfortably carry several at once, with his weapons slung over his back. They plant these stakes so closely together with their branches intertwined that it is impossible to tell which trunk each branch is attached to and vice versa. In addition, the stakes are so sharp and so closely interwoven that there is not enough room to insert a hand between them. Thus, it is impossible to grasp anything and pull it out, since the interwoven branches form a continuous chain. If by chance one of them is pulled out, it does not leave a large gap, and it is easily replaced.
>
> Livy 33.5

Various writers mention that soldiers carried bundles of wooden stakes with them. Scipio's soldiers at Numantia carried seven *valli* each, specifically for use in entrenching camps. But they did not always have them available and in Spain, Caesar's soldiers had to fetch materials for the stockade from a considerable distance. Wooden stakes, often erroneously called *pila muralia* (really a mural spear used for the defence of fortifications) have been found at a number of military sites, particularly the fort of Oberaden in Germany. The wooden stakes are up to six feet (2m) in length, with a waisted section, often called a 'hand-grip' half way down. Many are identified as belonging to a particular century or cohort. The stakes are in all probability the *valli* that Roman soldiers carried with them on campaign, a prefabricated version of the branches that Livy is describing, and more convenient to carry.[83]

There is a tendency to imagine these stakes as being stuck vertically into the rampart to create a small fence, but this design is problematic. Since both ends of the stakes were sharpened, it would have been very difficult to knock them into a rampart without the ends blunting or breaking. According to Pseudo-Hyginus, a palisade was not always put up in conjunction with a rampart, particularly if the army was camping on rocky or friable ground. This would certainly not have allowed the stakes to be knocked into the ground, so they must have been used in a different way. Vegetius supplies the answer: he states that the ramparts were defended by wooden *tribuli*, giant wooden caltrops, and this is very probably how the stakes were used. Three of the stakes could be tied together to form the caltrops, the waisted section in the middle being designed to allow them to be bound

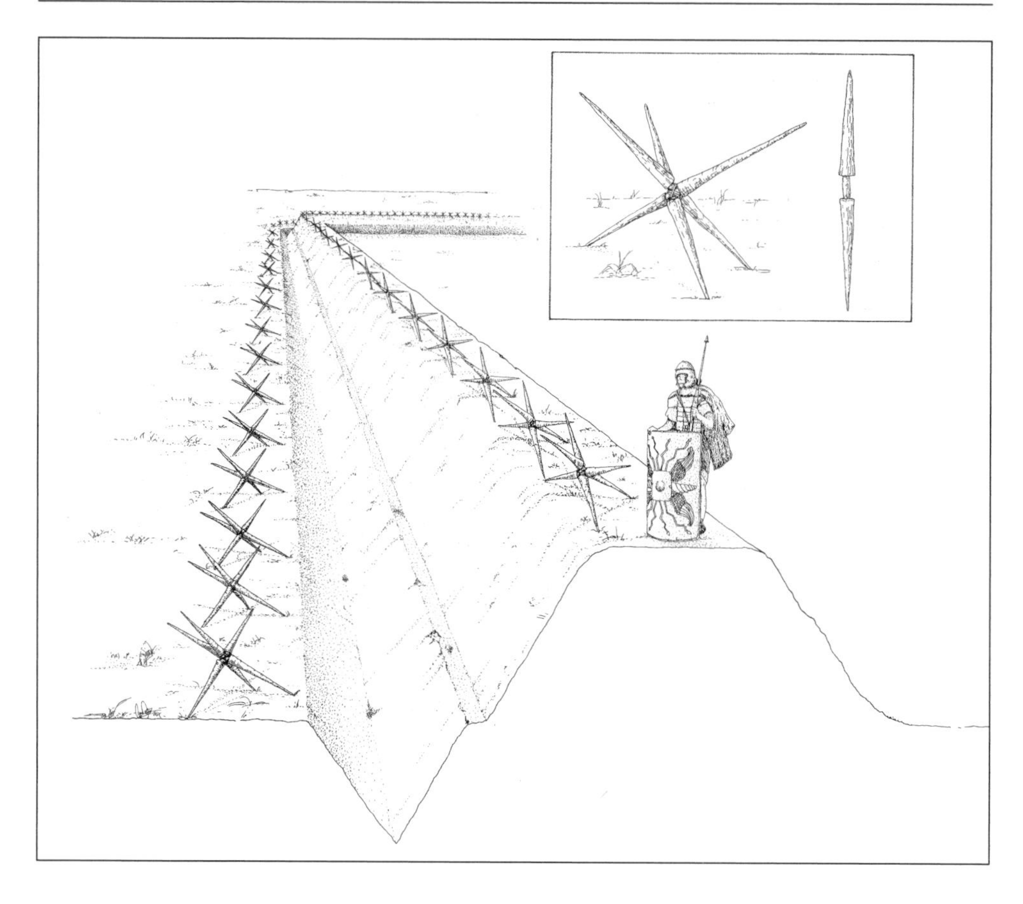

32 *Hypothetical reconstruction of marching camp defences* (Ian Dennis)

together more easily without slipping. Placed securely together on the rampart, these caltrops would create a spiky barricade like that described by Livy, having an effect similar to barbed wire. A formidable obstacle could be quickly and easily contrived, without the need for a rampart **(32)**. The finds of these stakes at permanent forts indicate their versatility in strengthening the defences of these forts as well as of marching camps.[84]

The gateways of marching camps were formed by gaps in the rampart. No towers were built, or doors placed in the gateways. These vulnerable entrances were protected by additional sections of rampart and ditch, the *clavicula* and the *titulum*. The *clavicula* (literally 'key') was an extension of the rampart and ditch following an arc around part of the gateway. *Claviculae* could be internal, external or double (in which case normally the rampart but not the ditch would be extended, on both the interior and exterior of the camp). The *clavicula* prevented attackers from charging straight inside the camp, and its design exposed the unshielded right side of an attacker to the defenders inside the fortifications of the camp. The most common type of clavicular gateway is the internal *clavicula*; there are only a few camps with external *claviculae*, whilst double *clavicula* camps are extremely rare (Troutbeck II and III in Cumbria are the only known examples in

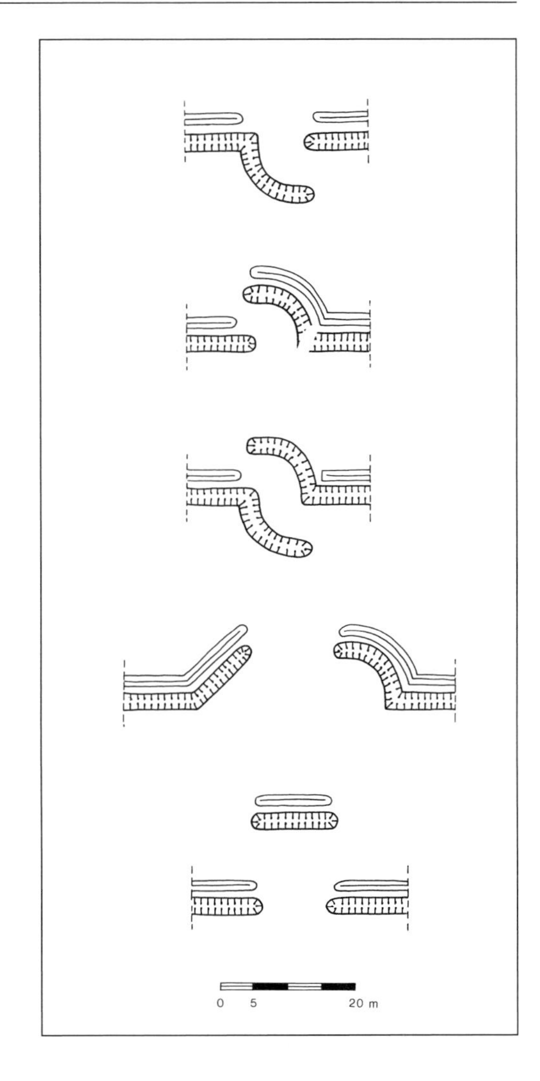

33 Gate types: from top, internal clavicula, *external* clavicula, *double* clavicula, *'Stracathro' type,* titulum

Britain). The *titulum* was a short section of rampart and ditch added some distance in front of the gateway. The word *titulum* may be a diminutive of Titus, the Latin generic for a soldier, much like the English 'Tommy', and like a soldier, it guarded the gateway. Pseudo-Hyginus recommends that it should be 60ft (20m) in front of the gate, but distances of about 30ft (10m) are more usual. Like the *clavicula*, the *titulum* prevented an enemy from rushing the gate head on **(33)**.

Pseudo-Hyginus recommends the use of internal *clavicula* and *titulum* at gateways, though this is a very rare combination in reality. Chapel Rigg and Glenwhelt Leazes marching camps in Northumberland have this combination, as does one of the practice camps at Llandrindod Common in Wales. The majority of camps have either an internal *clavicula* or a *titulum*. The two types of gateway defence were contemporaneous, though the *clavicula* seems to have fallen out of use after the Trajanic period, its depiction on Trajan's Column being some of the latest datable evidence for it. The military surveyor responsible for laying out the camp probably decided what type of gateway to use. The military situation may have been a factor in deciding the gate type, as may topographical concerns (in a confined space an internal *clavicula* might be the only option), and personal preference. Pseudo-Hyginus claims to be detailing a new method of camp organization and surveyors may well have had the freedom to try their own innovations. This may explain the variant of the *clavicula* gateway known as the 'Stracathro type', named after one of the marching camps in Scotland where it was used. Fourteen camps in Scotland have gates of this type, and all have been dated to the Flavian period. It is very possible that all the camps were designed by the same innovatory military surveyor.[85]

The setting up of a camp and construction of the defences would have taken a considerable time when the army stopped at the end of each day's march, possibly two to three hours or even more. The length of time would depend on the location, the type of terrain, the strength of fortifications required, and the proximity of the enemy. Surveyors were sent ahead of the army along with picked soldiers to choose an appropriate site to camp, and to mark out the lines of the camp with coloured flags. As Polybius says, this would allow the soldiers to get their bearings more easily and reduce confusion at this potentially vulnerable time. The actual construction of the fortifications however would have to wait until the bulk of the army arrived. As with the order of march, the number of times Roman armies were attacked whilst entrenching illustrates the vulnerability of an army engaged in this labour.[86] Tacitus explains why:

> Even entrenching camp was a formidable task in such close proximity to the enemy, for there was the danger that the men, scattered around and engaged in digging, would be thrown into disorder by a sudden sortie.
>
> *Histories* 3.26

If the army was entrenching in the presence of the enemy then, Vegetius states that:

> All the cavalry and that part of the infantry excused fatigues because of their rank stand guard in front of the ditch in an armed line and repel any attacks by the enemy.
>
> Vegetius 3.8

Screens such as these were frequently deployed when an army was entrenching in the presence of the enemy. When Aemilius Paullus entrenched his camp before the battle of Pydna in 168 BC, he had to do so in the presence of the enemy. His heavy infantry constructed the rampart and ditch behind a protective screen formed by the cavalry and light infantry. In the campaign against Ariovistus, Caesar had his troops entrench camp very near to the Germans' own encampment. He ordered the first two lines of his triple battle-line formation (the *triplex acies*) to protect the rear line that actually entrenched the camp. He did the same in Spain in the civil war campaign against Afranius. Once those camps had been fortified, the troops withdrew inside the defences to relative security. The skill of entrenching camp at speed was therefore an important one, and Vegetius states that learning to entrench a camp was one of the most important aspects of a recruit's training. In his speech to the African army, Hadrian indicates that camp building was a fairly

34 *Entrenching camp: the soldiers' arms are set up nearby in case of emergency.* Courtesy JCN Coulston

common military exercise, and practice camps are well known in Britain, particularly in Wales. Several training areas have been identified, including Gelligaer Common in South Wales, and the 18 practice camps at Llandrindod Common in Central Wales.[87]

Camp size and organization

> As far as possible, the camp should be 3 x 2 in proportion so that a blowing breeze can refresh the army ... for example, 2400 feet long by 1000 feet wide. If it is longer the trumpet call can be sounded, but in a disturbance the horn cannot be easily heard at the *porta decumana*; if it is wider, the outline is too near being a square.
>
> Pseudo-Hyginus 21

Vegetius agrees with this advice, though he adds that the camp should be shaped in accordance with the topography of the site. The majority of marching camps in Britain have proportions ranging from the 3:2 'playing card' shape recommended by the treatises to a much squarer shape, with the length only slightly more than the width. Camps and forts dating to the Flavian period tend to be of this latter type, the advantage being that a square can enclose a greater area than a rectangle despite the overall lengths of the perimeters being the same. Although topography and defensive requirements might affect the shape and proportions of the camp, the overall design, as with the gate defences, probably lay with the military surveyor.

The method of camp surveying and organization described by Pseudo-Hyginus is based upon the size of the army. Thus, the overall size of the camp should be proportionate to the number of men and supplies. Onasander and Vegetius agree with this principle. The main reason for this relationship between size of army and encampment is of course the defence of the perimeter when necessary. A camp could not be properly defended if there were too few soldiers to cover the perimeter, and a shorter perimeter would reduce the amount of time and effort spent in the potentially dangerous activity of entrenching camp. Marching camps then should in theory provide some information about the size of the armies that built them, and attempts have been made to trace the progress of particular campaigns from the distribution of camps of comparable size and design.[88]

As far as the size of camps and armies is concerned, the detailed figures provided by Pseudo-Hyginus can be used to calculate in theory the size of the army occupying a camp, but there are factors which complicate such calculations. Cavalry required more space than infantry, so the proportion of cavalry in the army would affect the size of the camp, as would the inclusion of 'dead space' such as marshy ground within the camp perimeter that could not actually be occupied. Another factor which might affect the density of troops in a marching camp is one favoured by the military strategists. Onasander advises generals to construct small camps and crowd their men in to deceive the enemy as to the size of the force. Frontinus includes the famous example of the consul Nero in the Second Punic War who joined his fellow consul by a secret march before the battle at the Metaurus in 207 BC. No increase was made in the size of the camp, though it was now very crowded, containing two armies instead of one. As a result of Nero's forced march and this deception, Hasdrubal faced a much larger Roman force than he was anticipating and was defeated. This type of ruse probably occurred rarely though.[89]

Pseudo-Hyginus gives such a clear explanation of his theory of castrametation that it is a comparatively straightforward task to reconstruct the army and camp that he describes in his work **(35)**. His army, some 41,000 strong, contained a mixed force of legions, auxiliary units and provincial levies, along with detachments of the Praetorian Guard, indicating an imperial campaign. Although some historians have suggested that Pseudo-Hyginus is describing an army that actually existed and campaigned on the Danube in the first or second century AD, it is unlikely that this is the case. It is instead probably only a hypothetical army, compiled on paper to illustrate how to encamp all, or nearly all, the different types of units a military surveyor might encounter. As a guide to camping Roman military units, Pseudo-Hyginus' work is fairly comprehensive, though with one obvious omission: he fails to indicate where the servants should camp. Camp followers such as merchants seem to have been excluded from camping within the fortifications, but they were not an official part of the army. The servants were part of the army, and presumably must have been allocated space within the camp. Pseudo-Hyginus, however, makes no mention of them, but it is likely that they shared the space allocated to each unit for its pack animals. This space allocation is generous, with each *contubernium* of eight men receiving an area 9 x 12 Roman feet for their *iumenta*, baggage and pack animals. Since the servants were responsible for the pack animals, it would seem logical for them to camp near to the animals.

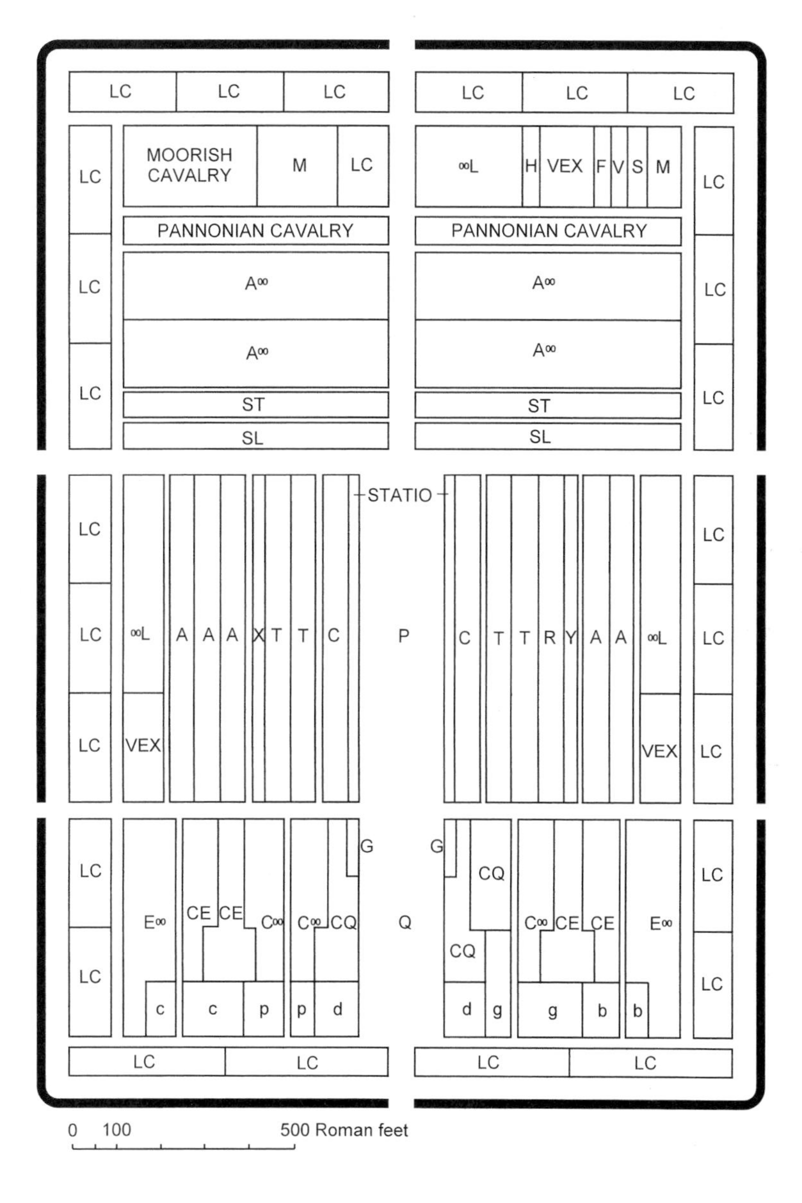

∞L	milliary legionary cohort	H	hospital
A∞	milliary ala	LC	legionary cohort
b	Britons	M	marines
c	Cantabri	P	praetorium
C	comites	Q	quaestorium
C∞	milliary infantry cohort	S	scouts
CE	part-mounted cohort	T	praetorian cohort
CQ	quingenary cohort	V	veterinary
d	Dacians	VEX	legionary vexillation
E∞	milliary part-mounted cohort	X	225 praetorian cavalry
f	fabrica	Y	225 equites singulares
g	Gaetuli		

35 *Reconstruction of Pseudo-Hyginus' marching camp*

36 *Marching camp at Rey Cross, Co. Durham.*
Courtesy Cambridge Committee for Aerial Photography

Because the basic internal arrangements of many surviving marching camps can be traced through the extension of the interior roads from the gateways, it is possible to impose Pseudo-Hyginus' methodology on camps, and attempt a hypothetical reconstruction of both individual camps and the armies that built them.

Rey Cross **(36)**

This camp, possibly dating to the campaigns of Cerialis in the early 70s, is one of a line of marching camps across the Stainmore Pass, from the Vale of York to the Eden Valley and Carlisle. The defences enclose some 18.6 acres (7.53ha), and nine gates survive, though there were probably originally eleven. If Pseudo-Hyginus' methodology is applied, the camp could hold the following units **(37)**:

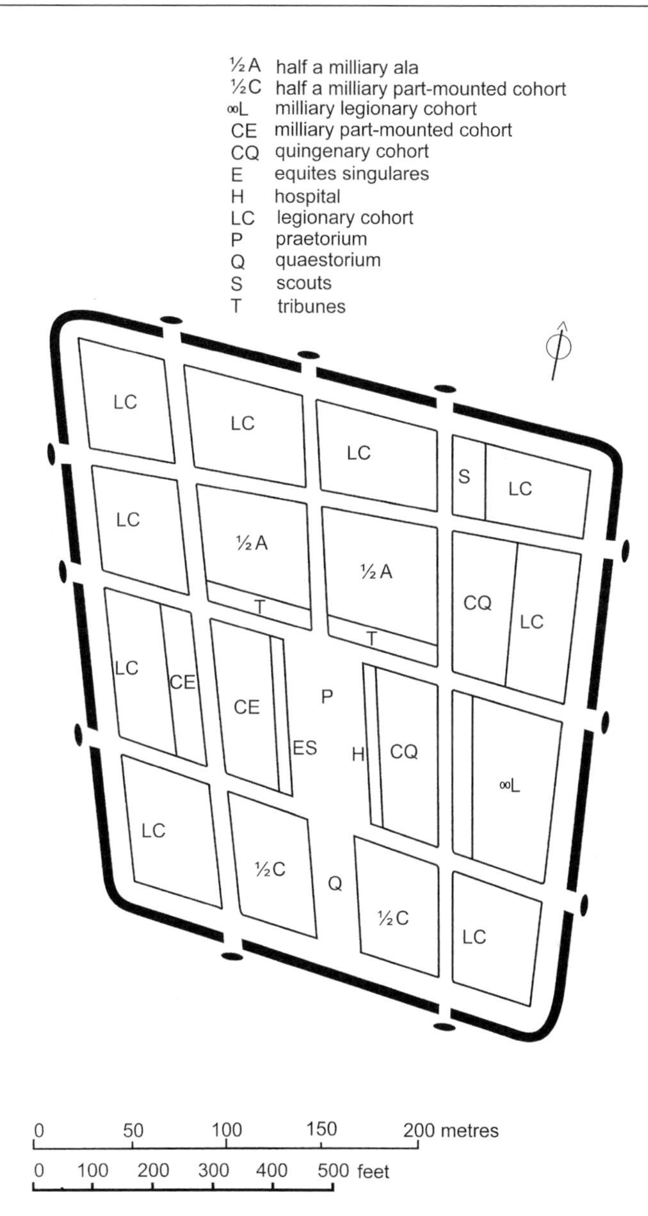

37 *Hypothetical reconstruction of Rey Cross marching camp*

	Infantry	Cavalry	
1 legion with a milliary first cohort	5,120		
1 milliary *ala*		720	
1 milliary part-mounted cohort	800	240	
1 quingenary part-mounted cohort	480	120	
2 quingenary infantry cohorts	960		
Scouts		200	
Total:	7,360	1,280	8,640

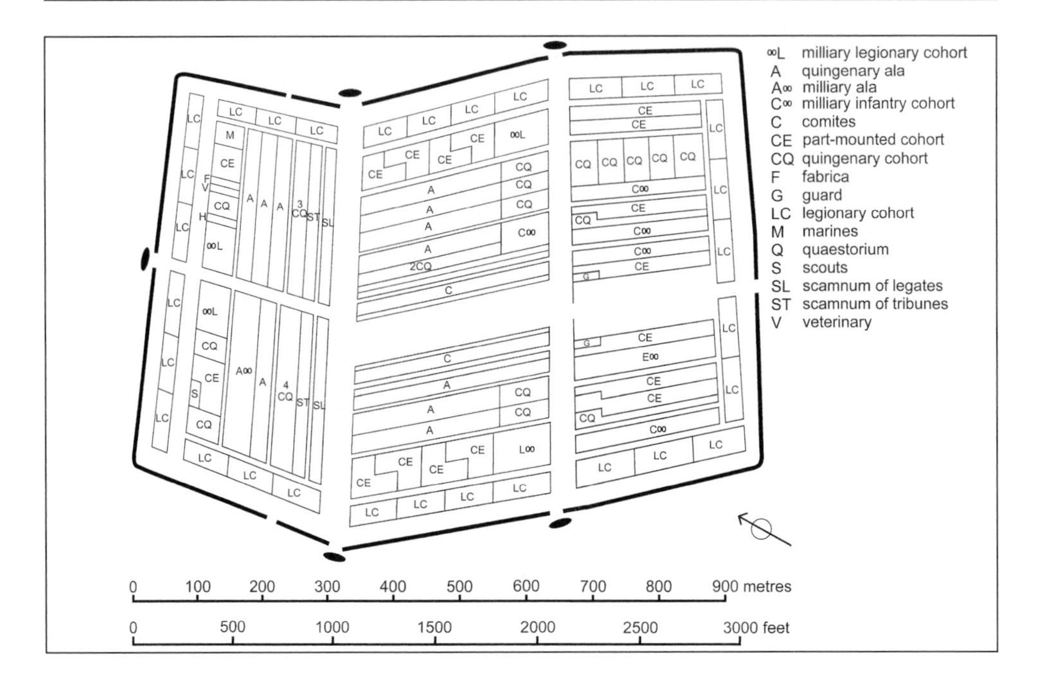

38 Hypothetical reconstruction of Durno marching camp, Grampian, Scotland

Durno **(38)**

Durno is one of the largest temporary camps in Britain at 143 acres (58ha) considerably larger than Pseudo-Hyginus' camp of 86.3 acres (34.9ha) which was designed for a hypothetical army of some 41,000 soldiers. It represents what has been described as 'a concentration of almost overwhelming force', leading to the identification of Durno as the possible site of Mons Graupius.[90] The camp is some 20 miles (32km) north-west of Aberdeen, one of a series of camps of at least 100 acres (40ha) in size stretching along the lowlands of Scotland north of the river Tay. By Pseudo-Hyginus' methodology, the camp could have contained an army of more than 53,000 men:

	Infantry	Cavalry	
2 legions with a milliary first cohort	10,240		
vexillation of 15 legionary cohorts including 2 milliary first cohorts	7840		
1 milliary *ala*		720	
11 quingenary *alae*		5632	
1 milliary part-mounted cohort	800	240	
17 quingenary part-mounted cohorts	8160	2040	
5 milliary infantry cohorts	4,000		
25 quingenary infantry cohorts	12,000		
Marines	600		
Scouts		200	
Guards	160		
equites singulares		600	
pedites singulares	300		
Total	44,100	9432	53,532

According to these calculations, Durno could have held almost the entire garrison of Roman Britain, and the hypothetical army of these calculations is that of a large imperial campaign. The only campaign of such proportions conducted in Britain was that of Septimius Severus in the early third century AD. Durno is certainly too large to have contained the 20,000 or so troops that Agricola deployed at Mons Graupius and probably does date to the Severan campaigns.

It is important to remember that these two hypothetical armies were invented for this exercise, and although most of these units could have been available in Britain, we know very little about the actual composition of campaigning armies in the province. The figures involved in these hypothetical reconstructions should also be taken as a maximum. In the case of Rey Cross, the troops are packed in to a density of c.460 per acre (c.1150 per hectare), near to the 475 per acre (1174 per hectare) recommended by Pseudo-Hyginus. In reality, some units may not have been at full strength, but to make matters easier the surveyor may nonetheless have assigned the 'standard' space for a unit of that type. Pseudo-Hyginus is describing what he calls a new method of camping, and it is impossible to tell whether this or some other system was used regularly by military surveyors. His theory can help to a certain extent with the understanding of camp design and the size of armies, but it must be used with caution.

Camps and campaigns

Groups of camps belonging to individual campaigns can be identified on the grounds of similar size and design. Their temporary occupation, however, makes dating them extremely difficult and so it is practically impossible to assign the different series of camps that have been traced to individual campaigns that have been recorded by ancient historians. Some estimates can be made about the approximate scale and date of campaigns. The camps over the Stainmore Pass, including Rey Cross, are all about 20 acres (8ha) in size, and represent the campaign of an army based on one legion with attached auxiliary forces, probably in the early Flavian period. The different series of camps in Scotland north of the Firth of Forth indicate armies of various sizes at different dates, including the imperial campaign of Septimius Severus in the early third century. These series also show how Roman campaigns in this area tended to concentrate on the lowland coastal areas, and failed to make serious incursions into the Highlands **(19)**. They do indicate, though, that the armies campaigning in this area could have been supplied partly by sea, and Tacitus' reports suggests that this was certainly the case during the later years of Agricola's governorship. He mentions the involvement of the fleet during the sixth season of campaigning, working in tandem with the army. Though he does not indicate any specific responsibilities for supplying the army, it seems likely that armies would have made use of water-borne supplies, whether by sea or inland water-ways, whenever possible.[91]

Conclusions

Although the idea of the campaign camp was developed well before Rome became a major power in the ancient world, it was the Romans who made the temporary camp a key part of their campaign strategy. They used the defended and organized encampment not only in siege warfare and before pitched battle like the armies of earlier civilizations, but on a regular, usually nightly basis to ensure a safe advance, no matter how hostile the territory. As in other aspects of campaigning, the Roman army's camping procedures varied according to the needs of the military situation. The location of camps depended on the proximity and nature of the enemy and, more importantly, the convenience of a water supply. Sometimes, possibly quite regularly if there was no immediate danger of attack, the camp's defences would be very slight, or non-existent, with security provided by lines of sentries. Some generals even risked such poorly defended encampments in the presence of the enemy, though are criticised for putting their armies in danger.

The presence of so many practice camps in Britain, and the stress placed in military textbooks on siting and surveying camps, indicates the importance that the Romans themselves placed on these skills and the routine of entrenching camp nightly. In pitched battle, as Livy reports in the speech of Paullus, the camp was of even greater importance, offering the potential for a reverse of the battle result. Control of the battlefield after a pitched battle was one indicator of victory, and if a defeated force was still present in an encampment, a clear victory might be denied.

4 Open Combat: pitched battle

Introduction

> Whoever sees fit to read these commentaries on the arts of war abridged from the most respected authors wishes to hear above all else the theory of pitched battle and instructions for fighting them.
>
> Vegetius 3.9

Pitched battle was seen in the ancient world as being the most likely means of achieving a decisive victory over the enemy, but also the one with potentially the most risk. Wars could be won or lost on the outcome of a single battle: Zama, Cynoscephalae, Pharsalus and Philippi were all decisive battles, having a major impact on the course or result of a war. The failures and successes in pitched battles during the Second Punic War saw the advantage shift from the Carthaginians to Rome, and a victorious conclusion for the latter which few could have expected in the aftermath of such heavy defeats as the Trebbia and Cannae. Roman generals might seek pitched battle to earn the glory of military success, perhaps the honour of finishing off a campaign, and the chance to triumph through the streets of Rome as a general hailed by his troops as *imperator*.

The big set-piece battle description seems to have been almost a requirement for any writer of Roman history, whether that was Livy compiling a grand history of the Republic or Tacitus writing a biography of his father-in-law, Agricola, and certain information was expected. The sizes of the armies, the dispositions, the speeches of the generals, the engagement, flight and casualty figures were all part of the description, though length and emphases might vary considerably. Cassius Dio's account of the Boudiccan revolt, for example, places far more emphasis on the speeches of the two leaders and the outlandish figure of Boudicca than on the actual events of the battle. Dio covers the revolt in twelve chapters, seven of which he devotes to the speeches of Boudicca and Suetonius Paulinus, whilst the pitched battle is all over in one. The speeches Tacitus gives to Calgacus and Agricola before Mons Graupius are over twice as long as the battle depiction. Appian may never have experienced battle himself, but his accounts of pitched battles are amongst the most graphic and detailed of all. On the other hand, Caesar's reports of his own battles can omit a lot of details, often leaving only the bare bones or the most significant events that took place.[92]

39 *Two metopes from Adamklissi depicting battle between Romans and Dacians; the legionaries use their short swords for stabbing and slashing.* Courtesy JCN Coulston

Despite the huge significance of pitched battle in the ancient mentality, there is little strategic discussion on a broader level in the military writers about whether or not a battle should be offered or accepted in the course of a campaign. Vegetius shows some awareness of the alternatives to pitched battle, such as skirmishing or destroying the enemy in an ambush, and Frontinus includes a very long collection of examples of ambushes in his *Stratagems*, but even so, the main emphasis in both these works is on pitched battle: they simply accept that pitched battle will occur during the course of a campaign. Precisely because battles were seen as being potentially so decisive, it was considered vital for the circumstances to be as favourable as possible. Failure to plan could result in a battle being lost, and though one could never plan against the influence of bad luck, there were many things that could be controlled to ensure a favourable outcome.

Preliminaries

> Good officers never engage in general actions unless induced by opportunity or obliged by necessity.
>
> Vegetius 3.25

All the writers of general military treatises show an awareness of the preliminaries to fighting a pitched battle. They provide not only lengthy advice on the disposition of forces for pitched battles, but also on the timing of battle, use of topographical features to the advantage of the general and other 'stratagems' of benefit to the commander. Both Onasander and Vegetius stress the importance of training for pitched battle. Onasander advised that soldiers should undertake mock battles on different types of terrain to prepare them for war, charging up hills to take imaginary positions and throwing clods of earth to simulate gathering and throwing missiles. Josephus suggests in his excursus on the Roman army that its success in war came from the training the soldiers underwent, to the extent that 'their drills are bloodless battles and their battles bloody drills'. It is a fairly common literary *topos* or theme for ancient historians to comment on how a general newly arrived in command of an army would quickly lick it into shape through an intensive training and drilling programme, an exercise that would have had considerable importance for the discipline and morale of the soldiers as well as their military skill.[93] Vegetius' comments on training are combined with perceptive remarks on the nature of warfare and pitched battle in particular that indicate that his sources understood these things well. He notes that soldiers who have not fought for a long time should be treated like new recruits. Both need to learn or relearn discipline, loyalty to their unit, unit cohesion and fighting skills.

> When legions, auxiliaries and cavalry arrive from different places, the best commander should have them trained by picked tribunes known for their industry in all types of arms, firstly in their individual units. Afterwards when they have been formed into one group he will train them himself regularly as if to fight in pitched battle. He will examine them to see what skill they will possess, what courage, how well they work together, whether they obey carefully the trumpets' orders, the directions of the standards and his own

> orders or command. If they fail in anything they should be exercised and trained for as long as it takes until they are perfect. But if they become properly skilled in drill, archery, throwing the javelin and drawing up the line of battle, even then they should not be led into pitched battle rashly, but on a carefully chosen occasion. Before that they should be 'blooded' in small-scale battles.
>
> Vegetius 3.9

Later Vegetius returns to the problems faced by new recruits or men who have not seen action for a long time. They are horrified by the sight of men being wounded or killed, confused by fear and begin to think of flight rather than fighting, though he does not mention the awful necessity in battle of having to climb over and trample on the bodies of dead comrades when advancing. Vegetius' caution about committing troops to the decisive engagement of pitched battle is further indicated when he suggests that before battle the soldiers should be allowed to become used to the enemy from a safe distance because they will then be less fearful.[94] The genuineness of these concerns and practicality of the solutions offered are well illustrated in the campaigns of Marius and Caesar in the late Republic. Marius took over command of the war against Jugurtha with a newly recruited army, which he took to a fertile area for the soldiers to plunder. He went on to attack some forts and villages that were poorly defended and offered little resistance, then fought some small-scale open battles. Sallust claims that this enabled the new recruits to learn the importance of remaining in rank and obeying orders, and gave an opportunity for recruits and veterans alike to meld into one cohesive unit. Easy victories with opportunities to plunder would also have encouraged the soldiers not to fear battle, and might have led to greater morale and confidence when they faced sterner challenges. The effectiveness of training and discipline is illustrated when the Nervii attacked Caesar's army as it was entrenching camp. The soldiers' military knowledge and experience from earlier battles meant that they knew what to do without being ordered, so could react quickly enough to form up ranks and recover after the surprise attack.[95]

Morale is of considerable importance to an army, and it was vital for the general to instil confidence in his troops before committing them to pitched battle. This could be done in a number of ways, including a speech before the battle started and fighting on foot with the infantry. But the army's morale could be boosted in other ways, even before the troops deployed. Onasander suggests parading cowardly and spiritless enemy prisoners through camp so the army will know they are facing a weak enemy (but advises the general to kill strong and courageous prisoners), and Frontinus has a collection of examples along this theme. Marius, as we have seen, built up the morale of his troops in Africa through cheap and easy victories, but used another approach in his campaign against the Cimbri and Teutones. He kept his army within strong camp fortifications whilst the enemy passed so they became used to the sight of the 'barbarian' Germans, became less fearful of them, and were eager to meet them in battle after their inactivity.[96]

The timing of battle

Frontinus provides a number of 'stratagems' concerning the timing of pitched battles. These examples illustrate delaying the deployment of troops to get the advantage over an enemy worn out from hunger and exposure to the elements, attacks when the enemy is hampered by religious scruples, and when the enemy is in the process of deploying his line of battle. Other military writers also suggest several of these tactics. All agree on the importance of an army being properly fed and watered before engaging in pitched battle, and therefore the value in forcing the enemy to engage when hungry or, as in the case of Tiberius when campaigning against the Pannonians, after the enemy had been formed up for battle all day exposed to fog and rain showers. The weather could be significant in another way too, since it could affect the use of some weapons. At Magnesia, the damp weather reduced the effectiveness of the Greek archers since the moisture had softened the bowstrings. Frontinus claims that it was a Roman stratagem to make use of the weather in this way, but it was probably just luck. The military writers all advise against fighting a pitched battle after a long march, and conversely suggest the enforcement of battle against an enemy tired out by a march. The Roman practice of entrenching a camp before battle ensured that usually armies paused, encamped and rested before offering pitched battle, but not always. Although accounts of campaigns in histories often contain examples of these kinds of 'stratagems', there were occasions where this kind of advice might be ignored. If the morale of the soldiers was particularly high, a general might engage his troops even though tired from the march. Both sides seem to have done this in the second battle of Cremona, though in this civil war the officers of neither side seem to have been able to control their troops at all successfully.[97]

Religious scruples might affect whether or not pitched battle might be offered or accepted and potentially influence the course of the whole campaign. Campaigns might be brought to a temporary halt because of a festival or a day unfavourable for military action. Roman armies were rarely hindered by religious scruples, but Greek and Jewish armies sometimes experienced considerable difficulties because of such concerns. According to Frontinus, Vespasian attacked the Jews on the Sabbath and so defeated them, and Pompey is reported to have done the same when he captured Jerusalem in 63 BC. Caesar too was aware of the advantages that could be gained from this kind of 'stratagem'. He mentions how he learned that the Germans under Ariovistus would not fight when the moon was waning, so he forced them to fight a pitched battle at that time. The Germans did not refuse to fight, but Caesar suggests that their fighting was affected by their belief that to do so was inauspicious, and this helped him gain a victory, for their morale must have been low.[98]

Frontinus also notes the advantage of offering battle towards the close of day so that retreat or flight could be made under cover of darkness. The example he uses is of Jugurtha's strategy against the Romans in North Africa. The Roman pursuit of the fleeing Britons after Mons Graupius continued only until night fell, when chasing a dispersed enemy would have been an impossibly difficult task, with a serious danger in the confusion of a night action of mistaking the identity of friends and enemies. The tactic recommended by Frontinus is valid, but of potentially more use to an army fighting a defensive campaign or one whose strengths did not lie in the line of battle. Given the defensive emphasis of Vegetius' manual, though, it is perhaps surprising that he does not suggest this stratagem in his section on pitched battle.[99]

Choice of terrain

> The good general should know that a large part of victory depends on the actual place in which the battle is fought. Be careful, therefore, to seize the advantage of terrain before you engage in battle...
>
> Vegetius 3.13

All the military writers stress the importance of the terrain where a pitched battle is to be fought, as they believed that victory was greatly dependent on the physical nature of the battlefield. The general is advised to choose terrain to suit the strengths and weaknesses of his troops, and those of the enemy. Since cavalry or infantry could have an advantage under particular topographical circumstances, choosing the right battlefield was of considerable importance. Holding the high ground also brought an advantage because weapons and missiles could be fired with greater force, and it is harder for troops climbing a hill to engage the enemy. Such is the importance of the field of battle that Onasander recommends avoiding battle until a suitable place is found. This was of course not always possible, particularly if the army was attacked on the march or if it was essential to force a pitched battle whatever the nature of the terrain. Even then, however, it was possible to take advantage of natural phenomena or even construct field-works to provide some kind of protection.

Although the military theorists clearly felt that the nature of the terrain was of fundamental importance in pitched battle, actual accounts of battles can be frustratingly limited in the amount of information they provide on this subject. It can be impossible to establish the exact site of a battle from the literary accounts and in fact none of the pitched battles that took place in Britain can be located with any certainty by using the literary accounts. Such details do not seem to have been of particular concern to historians, who may in any case have been uncertain themselves about the location. Topographic details outlining natural features like rivers and hills are often very vague, and when more details are provided, it may be because problems had arisen from the nature of the terrain. The very uneven and cut-up ground on which the First Battle of Cremona was fought caused difficulties to both sides, with units unable to advance in formation, losing contact with each other and causing vulnerable gaps to appear in the battle line. There are occasions though where historians are more forthcoming and include interesting details. The last major pitched battle of the civil wars between Caesar and the Pompeians was fought at Munda in Southern Spain, on a plain about five miles in extent; the plain was divided by a stream with a marshy bank. The Pompeians had deployed across the stream, ensuring that Caesar's men would have to cross it in order to engage, then press on uphill towards the Pompeian lines. The historian notes how advantageous the situation was for cavalry, since the terrain was fairly level and the weather calm and sunny. During the battle it was Caesar's cavalry that was able to take advantage of the terrain, attacking the Pompeians' left wing to prevent them sending reinforcements from there to their right which was being pressured by the infantry. Polybius' fascinating critique of Roman tactics in battle against the Insubrian Gauls in 223 BC includes severe criticism of the general Flaminius for deploying his army with their backs to a river, making the usual Roman tactic of maniples withdrawing through the battle line an impossibility.[100]

One major concern of Roman generals was in holding high ground in a pitched battle or in preparation for one. The two main advantages are pointed out by Vegetius: the extra height increased the range and effectiveness of artillery, and made it harder for the enemy to engage. It meant that an attacking army would be able to charge down on the enemy with greater force, and it was better also for an army on the defensive. Frontinus attributes Pompey's easy victory over Mithridates in 66 BC to the Roman deploying on a hill and charging down on the enemy with irresistible force. Tacitus describes the difficulties the Romans faced in an uphill attack at Mons Graupius, where Agricola had to throw in his reserves to keep up the momentum of the attack. When Caesar deployed with his artillery on high ground to fire on the Bellovaci, the Gauls had more sense than to engage under such circumstances. It was only an advantage to hold the high ground, then, if the enemy attacked despite the difficulty of the terrain. Agricola was confident enough to do this at Mons Graupius, or compelled to do so if he wished to fight the decisive battle he had been seeking for so long. Caesar's army was taken by surprise when entrenching at Zela, something that happened to him several times in the Gallic War. His excuse this time was that Pharnaces had deployed his army on a slope below Caesar's campsite and was trying to engage Caesar on ground so unfavourable to his own army, that no one could possibly have expected the Pontic king to commit his men to such a foolish charge. The narrator of the Alexandrian War claims that Caesar could not believe that Pharnaces intended to engage, it must have been a feint attack, so Caesar's troops were still entrenching when the attack came. Needless to say, the Pontic troops, having charged up a hill, were quickly routed, allowing Caesar to claim, *'veni, vidi, vici'*.[101]

Marshy ground was frequently unfavourable to Roman armies. None of the military writers warn about this, but it is quite clear from historical writings that the Roman legions in particular experienced difficulties when operating on this type of terrain. We have seen Caesar's concerns about the battlefield at Munda because to engage the enemy his troops had to negotiate a marshy river. Caesar seems to have been expecting the Pompeians to oppose him when crossing it, but fortunately for him they did not. During his conquest of Gaul, at the Aisne in 57 BC, Caesar had deployed on one side of a marshy river, against a Gallic army on the other **(40)**. Both sides waited to see if the other would dare to cross, ready to attack when those advancing were impeded by the terrain, but neither was prepared to cross the marsh. The Gauls broke the stalemate by attacking across the river Aisne, to the rear of the Romans, but were caught in difficulties in the river and suffered heavy losses. Tacitus describes well the problems experienced by Germanicus' army in Germany when trying to fight the tribes in the marshlands of the lower Rhine. The ground was too soft for the heavy armour of the Romans and they could not get a foothold to throw their missiles. The Cherusci, by contrast, used long spears suitable for fighting in this kind of terrain and, according to Tacitus, were long-limbed and naturally suited to marsh dwelling. Most valuable though would have been their experience in warfare on this type of terrain. When faced with an opposed river crossing at the battle of the 'Medway' in AD 43, Aulus Plautius made use of German auxiliaries who could swim across rivers in full armour. These were probably the Batavian auxiliaries who were also the first troops to cross the Thames as well as having important roles in the attacks on Anglesey by Suetonius Paulinus in AD 60 and later in the 80s by Agricola. Apart from

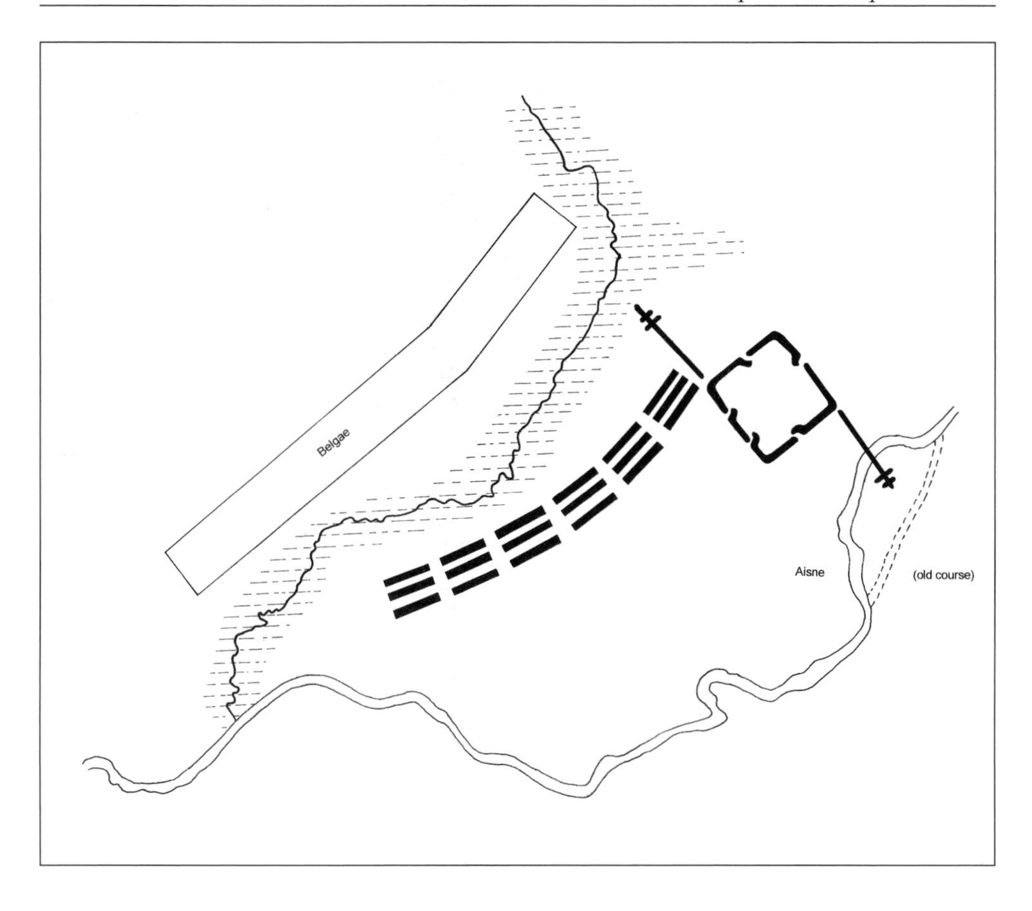

40 *Caesar's dispositions and fieldworks against the Belgae at the Aisne (after Kromayer and Veith)*

slingers and archers, the Roman army does not seem to have had units of infantry specially trained for particular fighting tasks, so the Batavians should not be seen as specialist 'amphibious' troops. A general, however, would be likely to make use of any particular abilities his soldiers might possess, such as local knowledge, or particular skill at crossing rivers.[102]

Rome's enemies knew as well as her generals what type of terrain would be unsuitable for Roman troops. On several occasions in the late Republic and early Empire, men who had previously campaigned for Rome as allies turned against her. Jugurtha in the late second century BC, Arminius and Tacfarinas in the early first century AD all turned to their knowledge of the Roman army and its strengths and weaknesses when fighting them, as did Julius Civilis in the Batavian Revolt of AD 70. This former auxiliary commander accepted pitched battle with the Romans under their commander Petilius Cerialis, but to give his Batavian army the advantage, Civilis dammed the Rhine to flood the battlefield. The Batavians were used to fighting on such terrain, giving them a big advantage over the heavily armed legionaries. There was no clear victor after battles on two consecutive days, though the Romans gained the advantage on the second. It was rare for an enemy army to withstand pitched battle with a Roman army so strongly in this period, and that was in no small part due to Civilis' stratagem of preparing the battlefield. It was also however, comparatively rare for Roman armies to be embarrassed so much by the terrain, and usually the variety of troop types in a Roman army would ensure the versatility needed for superiority on most types of terrain.[103]

In addition to making best use of the terrain, the general was also advised to take advantage of natural phenomena when considering his dispositions for battle. Vegetius recommended deploying so the enemy would get the sun, dust and wind in his face. According to some sources, this is what Hannibal did at Cannae, but Polybius adds that both sides deployed so that neither was put at a disadvantage by the rising sun. It is often impossible to tell from the accounts of battles whether a general took such factors into account or not when deploying. The literary accounts claim that Marius deployed his troops against the Cimbri and Teutones so that the Germans had to face the sun, and the wind blew dust into their faces. The story suits the literary tradition of Marius as a very shrewd and cunning general with a flair for stratagems, but in reality there were probably other factors affecting the dispositions of both sides that our sources fail to mention. It was only by unlucky coincidence, because the Flavian forces were arriving from the East, that the Vitellian soldiers found themselves facing the rising moon at the Second Battle of Cremona. This prevented them from judging the range of their catapults correctly, and made them perfect targets for the Flavian missiles.[104]

One of the best ways to take advantage of the terrain was to use natural obstacles to prevent outflanking manoeuvres. Such a preventative measure was particularly important as an army was at its most vulnerable when attacked on its flanks or at the rear. This had happened to the Macedonian phalanx at Cynoscephalae in 197 BC **(41)**, and to Pompey's forces at Pharsalus in 48 BC. The manoeuvrability of Roman maniples and cohorts aided both the execution of outflanking movements and their prevention. With these organizations it was possible for the rear maniples or cohorts in the line of battle to turn and meet an enemy coming from behind, something that was much harder for phalangists to do. However, despite this flexibility, it was better to prevent an attack on the flank or rear than to react to one, and both Onasander and Vegetius advise using natural features to secure one wing. The Roman left wing at Magnesia in 189 BC was secured by a river, a sensible plan since Scipio Asiagenes' army was seriously outnumbered **(42)**. Pompey did the same at Pharsalus with his right wing so he could concentrate his cavalry and light infantry on the left in the knowledge that the right could not be easily outflanked, protected as it was by the river Enipeus. Caesar was thereby forced to make alterations to his own dispositions in order to counter his opponent's strong left wing. This tactic could be taken further and natural features could be used to anchor both wings. Suetonius Paulinus, when greatly outnumbered by Boudicca's army in AD 60, positioned his line of battle in a steep sided valley to secure his wings, and with a wood at his rear to prevent outflanking movements, or an attack from the rear. Arrian intended to secure his wings on hills for his engagement against the Alans, but his strategy reduced his flexibility in that it would almost certainly involve fighting under quite particular topographical circumstances: Arrian mentions that the battle will be fought at an appointed place, but he does not reveal his intentions should he not find that place.[105]

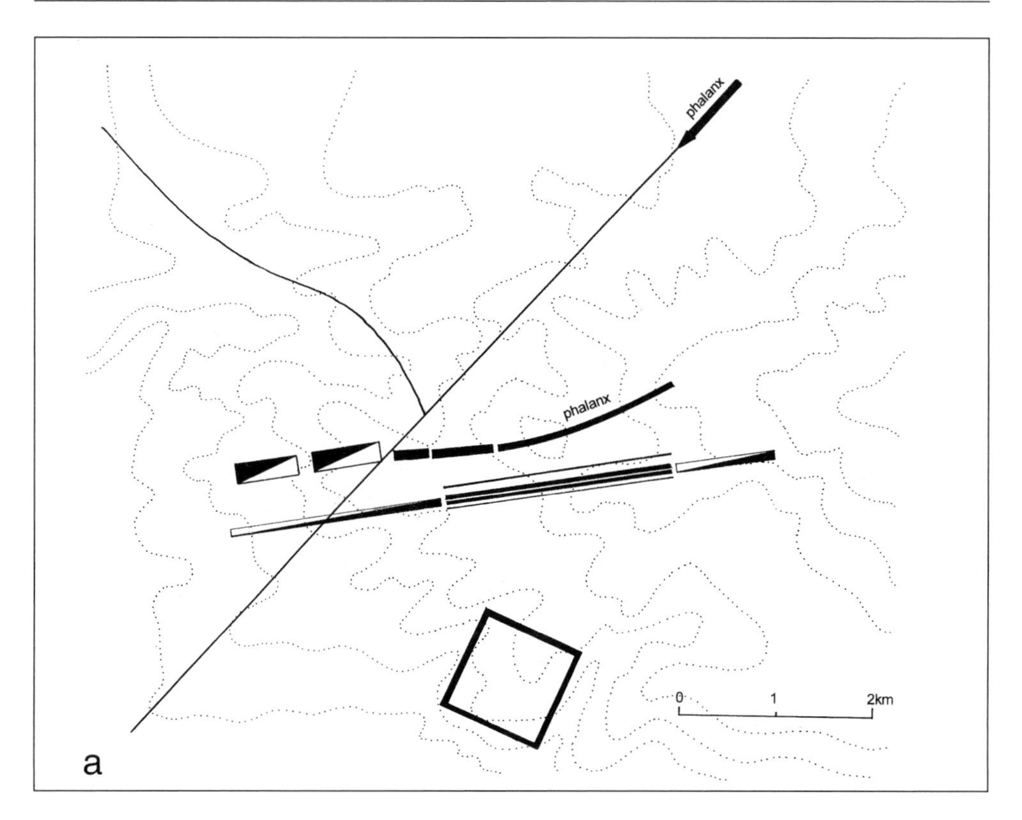

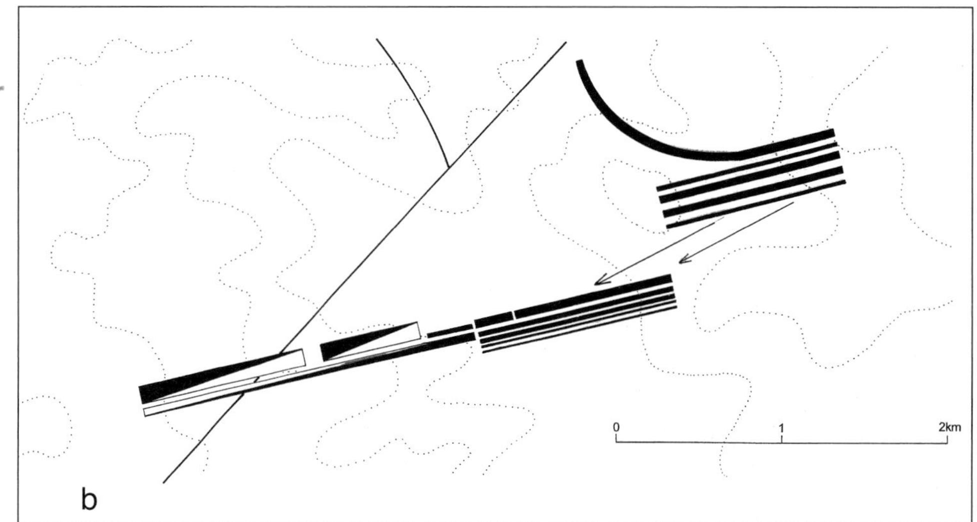

41 *Cynoscephalae. Philip was only able to bring half his Macedonian phalanx into action, but this pressed the Roman left hard (a). The Roman right was successful against the arriving left wing of the phalanx and pushed it back down the hill (b). One of the Roman tribunes then detached the* triarii *from the rear of the victorious right and charged it into the rear of the phalanx (after Connolly)*

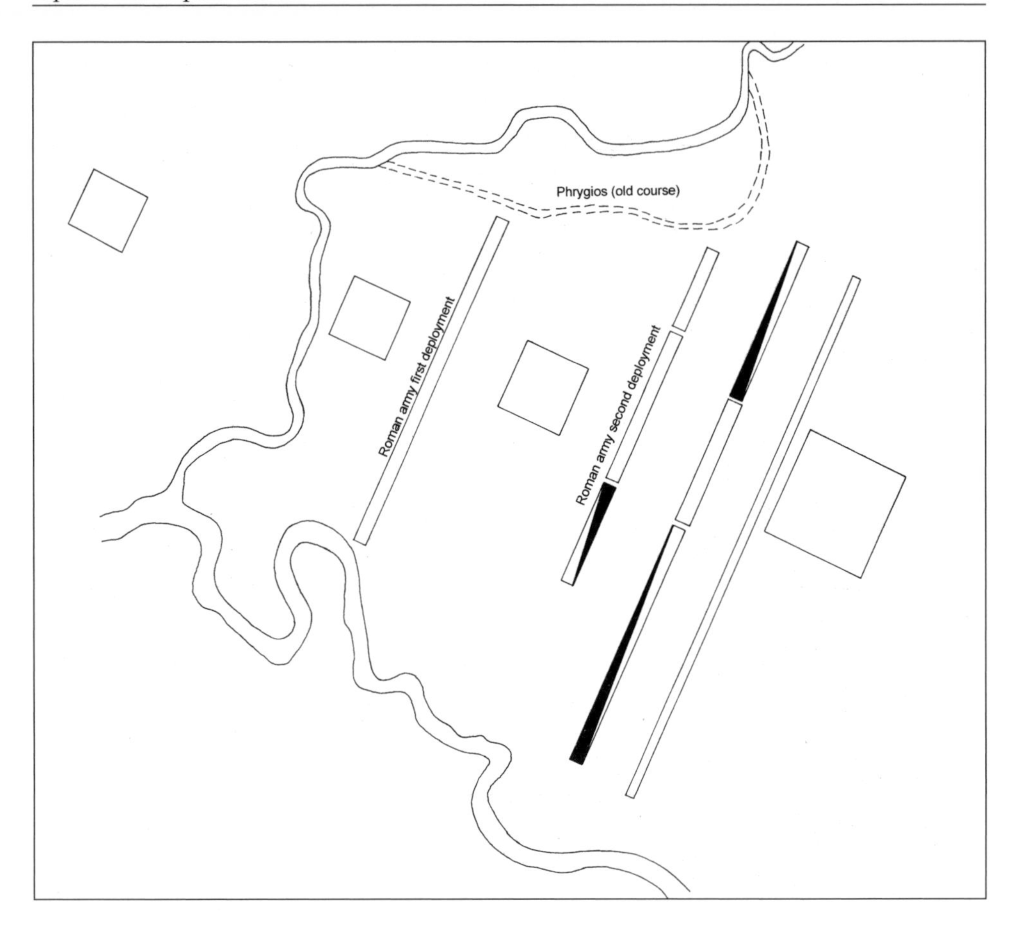

42 *Magnesia. Both sides attempted to anchor their wings on the rivers, and the Romans had built successive marching camps as both sides jostled for position before accepting pitched battle (after Kromayer and Veith)*

Fieldworks

When natural obstacles were not available, an alternative was to dig entrenchments to protect the flanks, a practice not mentioned by any of the military writers. This seems to have been a comparatively rare precaution, possibly because natural obstacles were preferable and with the potential for a lot of marching and counter-marching as two sides jostled for the best position for pitched battle, there was often no guarantee that battle would take place at a particular spot. Both Sulla and Caesar used trenches to protect their flanks in open battle, placing artillery at the end of the trenches for added protection. Caesar's trenches were four feet deep, probably enough to discourage the Belgae at the Aisne in 57 BC from attempting to cross them and attack his rear. Fieldworks Caesar had dug at Uzitta in 46 BC when attacking the town became a convenient obstacle on which to attach his right wing when facing Scipio in battle shortly afterwards. As when facing

Pompey at Pharsalus, he was then able to load more forces on his more vulnerable left wing. The disadvantages of fieldworks like these were that if the army advanced beyond them, the advantage was lost, and they could in any case prevent or hinder one's own army from carrying out out-flanking manoeuvres or reacting quickly to an unexpected situation that arose in battle. We may see a hint of this at Munda where, because of the unfavourable nature of the terrain, Caesar tells us he 'began to restrict the area of operations'. This may have involved digging trenches similar to those mentioned above (he does not tell us), but his men complained that this would hamper their chances of fighting a decisive battle. Any action that restricted the movement of troops on the battlefield, whether physical barriers or other means, reduced the flexibility of the army, one of Rome's principal advantages in fighting pitched battle, and this may have been why Caesar's troops at Munda complained.[106]

Final preparations

Vegetius includes little advice on the immediate preparations for fighting a battle, such as speeches, religious observances and the preparation of weapons, but Onasander does, and in his examples of pitched battles, Frontinus demonstrates an awareness of the importance of such matters. Onasander advised generals to ensure that their armies prepared and cleaned their equipment before battle. An impressive looking army, he says, will frighten the enemy more. This may well have been true, but the preparation of equipment had other significance for the soldier. The Roman army's system of rewards and punishments was aimed partly at the importance of collective responsibility of the unit as a whole. The preparation of equipment then, could help to encourage loyalty to one's unit and fellow soldiers, and a resolve not to let them down in battle. Equipment, particularly shield designs, could also identify an individual unit, and the brave (or cowardly) actions of a unit or soldier on the battlefield. Caesar's soldiers had highly decorated equipment which, according to Suetonius, ensured they looked impressive on parade and would be reluctant to lose their kit in battle. Loss of equipment on the battlefield could bring shame, and Frontinus cites the example of Cato the Censor's son who at Pydna in 168 BC went back among the enemy to retrieve his lost sword. Seeking a weapon was one of the only reasons for leaving ranks covered in the military oath soldiers swore, and presumably this covered leaving ranks both on the march and in battle, as Cato did.[107]

Religion also played its part in the preliminaries to battle. In the campaign camp,

> The altars are set up at the end of the praetorium; we will place the *auguratorium* to the right side of the praetorium next to the via Principalis, so that the general can observe the omens correctly; the tribunal is set up on the left side, so that having observed the omens, the general can ascend the tribunal and address the army on the favourable auspices.
>
> Pseudo-Hyginus 11

Onasander expected the general to take the auspices before leading his men into battle, in the belief that favourable omens would ensure they fought with greater courage and confidence. He also refers to the practice of taking and re-taking the auspices until they are favourable. Whatever the religious beliefs of commander and soldiers, and whether or not they believed in the omens, the procedure itself will have been part of the preparation for battle, like the readying of equipment. The involvement of the whole army in the religious procedures prior to battle would have encouraged loyalty to commander and fellow-soldiers, and may well have increased the confidence of the army in the case of favourable auspices.[108]

Frontinus illustrates the importance of favourable omens to the morale of the army in his collection of stratagems on dispelling the fears caused by unfavourable omens. These are not concerned with the auspices taken by the general in the presence of the army, but rather unexpected events. These might at first appear to be unfavourable omens, but a quick-thinking commander might re-interpret them and come up with a more favourable explanation. When Scipio Africanus tripped and fell on landing in Africa, his soldiers were horrified by what they took to be a bad omen, but he encouraged them by claiming that he had 'hit Africa hard'. Natural phenomena such as earthquakes, meteors and particularly eclipses could also be interpreted as bad or good omens. An eclipse of the moon the night before Pydna in 168 BC was interpreted by the Macedonians as a bad omen, signifying the eclipse of their nation, and this may have affected their morale in battle the following day. According to Livy, however, the Romans had predicted the eclipse, and its timing and cause had been explained by the officers to their soldiers. They did not then see it as an omen, and were impressed by the knowledge of their officers. It is noteworthy, however, that the military writers do not advocate ignoring the formal auspices taken before battle, or deliberately misleading soldiers about their meaning. This did happen on Roman campaigns, most famously at the naval battle at Drepana in 249 BC where the consul Publius Claudius Pulcher was not prepared to wait for the sacred chickens to eat their grain, the sign of a favourable outcome. Claudius had them thrown overboard instead and got on with the battle, a significant defeat for the Romans. Other Roman defeats were blamed on an error or failure in the conduct of religious observances prior to battle, including the major defeats at Cannae in 216 BC and Arausio in 105 BC. This explanation for a defeat deflected blame from military failings and may on occasion have helped to limit the negative affects on army morale in future engagements.[109]

Speeches

There was an expectation among military writers and historians that speeches would be made by the opposing generals to their troops before battle. This is why Onasander includes oratorical skills in his rather exacting list of the qualities necessary for successful generalship. The speeches recorded in histories are essentially rhetorical exercises that tell us far more about the traditions of writing history in antiquity than about what generals actually said to encourage their troops before battle. It is clear that generals did address their troops before battle, either in an assembly in camp after taking the auspices, as Pseudo-Hyginus suggests happened, or, perhaps more often, to troops on the battlefield after deployment. At one point in his Gallic Wars, Caesar lists all the things a general should do before battle: hoisting the signal; ordering the men to arms; sounding the trumpet; recalling the men from fatigues; deploying the line; encouraging the soldiers with a speech; and giving the signal for battle. When attacked by the Nervii, however, there was no time to do any of these things. Although Caesar dashed round the various sectors of his battle line to harangue his soldiers, most had already engaged the enemy and any speech of encouragement was unnecessary. At Pharsalus, where both sides deployed more slowly, Caesar mentions that he addressed his soldiers before the battle to encourage them, 'in accordance with military custom'.[110]

Practical considerations may have demanded that brief words of encouragement were addressed to different sectors of the battle line, as Caesar aimed to do in the attack by the Nervii, rather than a general address to the entire army. A general address would probably not have been audible to more than a small proportion of the battle line stationed near to the general. Hadrian's address or *adlocutio* to the army in Africa was given individually to the different segments of the army, in this case, the legionary centurions and cavalry, the cavalry ala, and the mixed cohorts, and the words chosen were appropriate to each group. The general on the battlefield would probably have done something similar. It is not inconceivable though that generals may sometimes have addressed the entire army, particularly if there was not much time to ride around the ranks. Even if most could not hear him, they could still have seen him and drawn encouragement from that sight, as soldiers did when seeing their general in the thick of battle.[111]

Disposition of forces

The line of battle or *acies* was the means by which an enemy might be broken. The vast majority of casualties in any pitched battle occurred when one *acies* was broken and had turned to flee, or if it was attacked on the flanks or in the rear. Because the outcome of pitched battle was seen as dependent on the strength of the *acies*, it is not surprising that the military writers spend considerable time explaining the organization of troops on the battlefield. Frontinus gives more examples of battle dispositions than almost any other type of stratagem, and on several occasions he explains the dispositions of both sides, and how one commander deployed his troops in reaction to the other. Most of Frontinus' Roman examples are of battles well known in the histories of Polybius and Livy, or the commentaries of Caesar, so there is little new information. The value of the examples lies in the emphasis on envelopment and flank attacks, and the value of light-armed and missile troops. The military writers mention the importance of planning battle dispositions in advance, and doing so with great care since, as Vegetius says, even with the best soldiers, a bad battle line will be broken up. Vegetius' advice is to deploy troops in their battle line first, and then wait for the enemy to deploy, so that there will be no interference from the enemy during the vulnerable period of manoeuvring. This advice may be indicative of inexperienced or weak Roman forces in the late Empire, for such matters do not seem to have concerned earlier writers. Onasander, perhaps more shrewdly, points out the advantage in waiting for the enemy to deploy first so the general can make his dispositions taking into account those of the enemy. This method would require considerable discipline and organization on the part of the soldiers since attack on a deploying army was a legitimate stratagem. The advice may be more relevant to better-trained troops of earlier periods. Advance planning and accurate intelligence were therefore important parts of the preliminaries to battle. Caesar frequently held councils of war with his legates, tribunes and senior centurions, and Arrian's *ektaxis* indicates that he had planned not just the formation of his battle line but his entire campaign very carefully.[112]

Whatever the differing theoretical views, in reality the question of when to deploy troops in relation to the enemy varied, as Onasander points out, depending on circumstances. Lucullus, greatly outnumbered by the army of Mithridates, was able to deploy quickly and attacked the huge and unwieldy Pontic army whilst it was still trying to deploy. The enemy line was in chaos and fled almost immediately. That there was less danger to disciplined troops when this happened is suggested by the account of Caesar's defeat of Pharnaces at Zela in 47 BC. The Romans were attacked whilst entrenching and Pharnaces' chariots charged. Even though the Romans had not properly deployed, they still repulsed the chariots with a volley of missiles, and seem to have been fully deployed to face the enemy infantry. The advantage of waiting for the enemy to deploy first though is well illustrated at Pharsalus in 48 BC. On studying Pompey's dispositions, Caesar saw that his enemy had posted all his light troops and cavalry on his left wing. Since this seriously threatened his own right wing, Caesar took one cohort from each legion's third line and posted them all to his right, reversing the advantage **(43)**. Scipio, on the other hand, deployed his army in the same manner against Hasdrubal in Spain in 206 BC each day until the day of the actual battle. The Carthaginian, expecting the same formation, deployed his troops accordingly, but Scipio had switched his dispositions suddenly, leaving Hasdrubal's army facing the wrong kind of troops for their specialisms.[113]

The theoretical writers agree that the battle formation depended on the type of soldiers the general had and how they were armed, and the strength of the enemy. In deployment, a balance had to be achieved between a long, thin line that the enemy could burst through, and a compact but deeper one that could be more easily outflanked. This was also an important consideration in the organization of the line of march, partly because of the vulnerability of that formation, but also because, as we have seen, an army's marching formation might be very similar or identical to its intended battle formation. Onasander's suggestions for the arrangement of the battle line are rather general: cavalry should face enemy cavalry (presumably on the wings), light troops posted in front of the main body of infantry should provoke the enemy to battle, then retire through intervals left within the ranks of the heavy infantry. This is very similar to the descriptions of Republican battle arrangements and tactics by Polybius and Livy, and Vegetius' recommendations are very similar. It is very likely, indeed, that Vegetius was using a Republican source for this section of his treatise, possibly Cato's now lost *de Re Militari*. Vegetius gives more detail than Onasander though, including seven prescribed battle arrangements, each designed for particular circumstances, strengths and weaknesses, and he claims that these would be sufficient to cover all possible scenarios. Both Onasander and Vegetius note that reserves should be kept at the rear of the main force to strengthen an endangered wing and to prevent flank attacks, but also for carrying out attacks on the enemy flanks and rear. Vegetius, as usual more elaborate and prescriptive than the earlier writers, suggests three different groups of reserves: on the right to attack the enemy left flank, in the centre to strengthen any weak part of the battle line, and on the left wing to prevent it (the more vulnerable wing) from being surrounded.

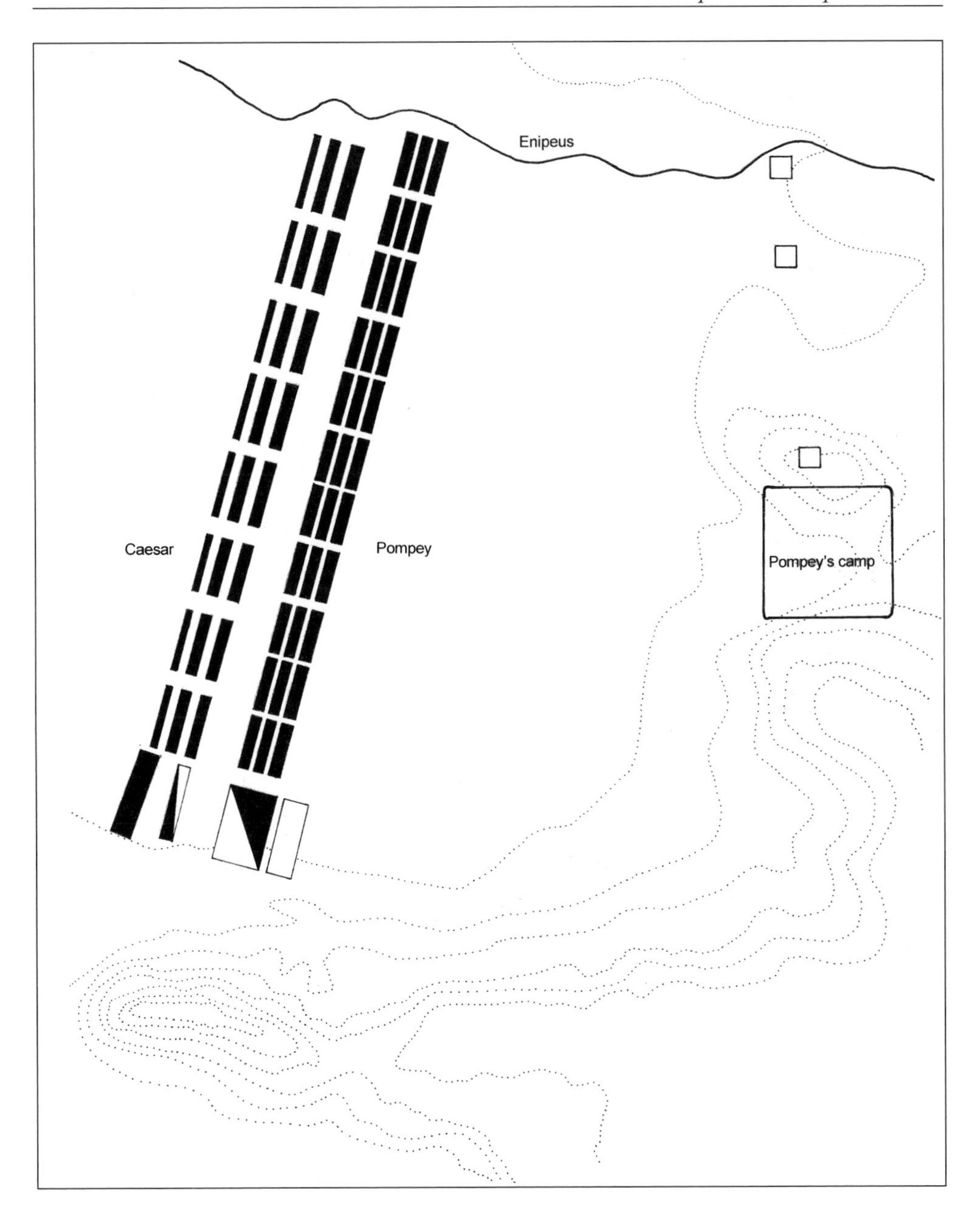

43 *Pharsalus. The armies of both Caesar and Pompey used natural features to secure their wings. The decisive actions took place where both generals had piled up their forces on Caesar's right (after Kromayer and Veith)*

Concern about the length and width of the battle line is best illustrated by Agricola's dispositions at Mons Graupius. According to Tacitus, Agricola was concerned about being outflanked so he 'opened out the ranks', increasing the gaps between the men. There were concerns that the line was too thin as a result of the manoeuvre, which would have created a less compact battle line, but this does not seem to have been a weakness during the actual battle. If necessary, as we have seen, fieldworks or a natural obstacle might have been used to protect one or even both flanks, and if a line was outflanked, reserves might be available to deal with the emergency.

We have already seen the basic similarities between the dispositions advised by the treatises and many of the battle lines described by Polybius and Livy. Polybius, indeed, considers the 'normal' disposition of the Roman acies to be as follows:

1 A screen of light infantry (the *velites*) to provoke the enemy to battle, and deal with any enemy light infantry or missile troops. They then retired through the ranks of the heavy infantry (exactly how is not known). The *velites*, being the most mobile of the infantry, would also assist the cavalry in pursuing a fleeing enemy.
2 The heavy infantry would take up the fight in their three lines of *hastati*, *principes* and *triarii*. The latter acted as a reserve force in the rear, but could also be used for outflanking manoeuvres as happened at Cynoscephalae in 197 BC.
3 The *socii* were stationed between the legions and cavalry. Some may have employed the same organization as the Roman legions, but others may have been armed differently.
4 The cavalry on the wings opposed the enemy cavalry, prevented outflanking manoeuvres and carried out the pursuit.

This is the basic formation regularly employed by Roman armies throughout the middle and late Republic. The adoption of a tactical system based on the cohort instead of the maniple in the late Republic does not seem to have affected greatly the basic dispositions of the battle line. There were no longer *velites* at the front of the battle line, but many of their duties as 'light troops', skirmishing, and providing missile fire and assistance in the pursuit, were taken by the non-Italian allies, the *auxilia*. Auxiliaries were also regularly stationed in the battle line between the legions and the cavalry on the wings, as the *socii* had been before the grants of Roman citizenship meant they could now serve in the legions. We do not see so often the deployment of light infantry in advance of the heavy infantry of the main battle line as the *velites* had been, though slingers and archers might be deployed there, or in the centre of the battle line, to provide additional fire-power. By the early Empire there appear to have been two basic types of disposition in regular use **(44)**. The first was essentially the continuation of the late Republican system described above with legions in the centre of the line flanked by auxiliary infantry with auxiliary cavalry on the wings. In the second arrangement the auxiliary infantry was deployed in front of the legions, with the cavalry on the wings. This is not a return to the system of *velites* with light troops beginning the battle and then retiring. These auxiliaries acted as the principal striking force of the battle line and the legions were held in reserve.

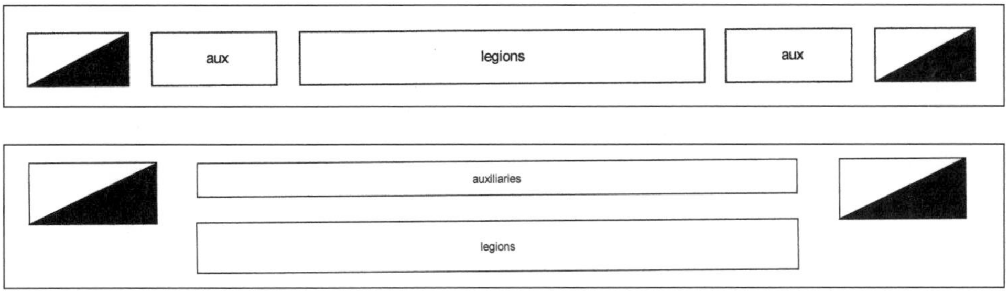

44 *Typical dispositions of the early imperial period: above, the continuation of the Republican system, and below, the use of auxiliary infantry as the main striking force*

The legion in battle

Although the cohortal legion lacked the differently armed troops whose deployment and role in battle Polybius and Livy describe, it was extremely flexible in the way it could be formed up in battle line. Its organization provided almost automatically reserves at the rear of the *acies* which could be used to strengthen a weakened section of the line, carry out outflanking manoeuvres, or deal with an attack on the rear of their own line. Vegetius suggests that the ten cohorts of the legion should be deployed in two lines of five cohorts, a *duplex acies*. Histories of the late Republic and early Empire rarely provide detailed explanations of the width and depth of battle lines, but Caesar and his continuators fortunately do. Caesar usually formed his legions in a *triplex acies*, with the cohorts arranged in a 4-3-3 formation. This arrangement probably evolved directly from the deployment of the manipular legion with its three different lines of troops, except that the soldiers were now uniformly armed. The *triplex acies* provided a reasonable compromise between the line being too short and it being too thin. The middle line of cohorts would have acted as a reserve for the cohorts in the front line, whilst the rear line could carry out outflanking manoeuvres or, if necessary, turn round to face an enemy attacking from the rear. We have already seen how at Pharsalus Caesar took a number of cohorts from the rear of his *triplex acies* to reinforce his right wing, whilst at Uzitta in 46 BC he created a fourth line of legionary cohorts on each of his wings to face Scipio's elephants. The very deep formation of infantry on the wings, together with slingers and archers on the right wing, forced the elephants to turn and flee, causing havoc with the remainder of Scipios formations.[114]

The flexibility of the cohort is also illustrated in Caesar's campaigns in Gaul. In battle against the Helvetii in 58 BC, Caesar's exposed right flank came under attack from the Boii and Tulingi who were allied to the Helvetii. In response, the cohorts in the rear line turned round to engage this new threat and the Romans fought a battle on two fronts. Later in the same year in battle against Ariovistus, when the Roman right came under pressure from the Germans, Publius Crassus, who was in command of the cavalry and could see what was happening, ordered in the cohorts of the third line. This manoeuvre swung the battle in the Romans' favour. This flexibility can be seen in the manipular legion as well, most famously at Cynoscephalae in 197 BC when an unnamed military

45 *Mobile artillery pieces could have added significantly to the fire-power of an army in pitched battle, Trajan's Column.* Courtesy JCN Coulston

tribune detached twenty maniples from the rear of the successful right wing and tore into the unprotected rear of Philip's phalanx, causing terrible carnage. It is this very flexibility of the Roman fighting organization that Polybius credits with Rome's successes over the Macedonian phalanx.[115]

Auxiliaries and missile troops

Throughout his work, Vegetius places considerable emphasis on the legion. Associating the non-Roman auxiliaries of the late Republic and early Empire with the 'barbarian' allies of his own day, whom he saw as poorly trained, ill-disciplined and untrustworthy, the fourth-century writer barely considers the importance of these troops. Onasander, on the other hand, gives particular prominence to the role of the light troops drawn from units other than the legions, and stresses their flexibility in battle. Light troops, he points out, were particularly useful in broken terrain and in dislodging enemy forces holding high ground. Slingers and archers did provide a valuable addition to the missile strength of Roman armies, and being lightly armed were more mobile than the legionaries with their *pila* were. Rome's light missile troops had traditionally been provided through treaty, alliance or some financial arrangement. The Balearics were famous for slingers, Crete for archers, but as Rome's empire expanded, so did the recruiting grounds, particularly for archers from the Eastern Mediterranean and Syria. Other auxiliary infantry, as discussed in Chapter 1, might be recruited locally for a specific campaign during the Republic. Once the *velites* of the legions had been phased out in the late Republic though, these auxiliaries came to have an increasingly important role supplementing the legions.

However, the general impression gained from the histories of the late Republic is that the fighting abilities of many of these auxiliaries were considered suspect. Appian states that the auxiliaries at Pharsalus were more for show than for use, and Caesar informs us that Crassus placed his auxiliaries in the centre of his battle line between the legions when fighting the Aquitani because he lacked confidence in them. Domitius Calvinus similarly placed the 'legions' supplied by Deiotarus, the king of Galatia, in the centre of his battle line against Pharnaces, with a small frontage, because he lacked trust in their strength. The auxiliaries were expected to fight more strongly and be more reliable if closely supported by the legions in this way. By being stationed in the centre of the battle line, these auxiliaries would have been protected on their flanks by regular legionaries and were less likely to be panicked by a flank attack. In the event of such an attack, the auxiliaries would have found flight a much harder prospect, hemmed as they were into the battle line by the legions. Caesar sometimes employed his auxiliaries as he did his recently recruited legions, in whom he seems also to have lacked confidence at times. These legions and auxiliaries regularly guarded camp during battle or were positioned towards the rear of the marching column. These auxiliaries though were not the regular auxiliary units of the Imperial period but were provided by pro-Roman leaders or under treaty, usually only for a particular campaign. With the regularization of auxiliary units during the early Empire and their establishment on a permanent footing, they came to provide a reliable and versatile part of the Roman military establishment, as we will see below.[116]

Apart from the *pila* of the legionaries, missiles were provided by auxiliary slingers and archers, and by artillery. Onasander suggests two positions for the slingers and archers: in the front of the line to fire directly at the enemy line, and on the wings to carry out flank attacks (where they could also protect their own flanks from similar attacks by the enemy). He notes how a missile attack on an army's flank would force the enemy lines together and cause them to become confused as they tried to avoid the missiles. We have already seen how at Uzitta in 46 BC, the archers Caesar placed on his wings were effective against Scipio's elephants. Scipio too had placed missile troops, in this case light javelin troops, on his right wing in a vain attempt to outflank Caesar and throw his lines into confusion with missiles. At Pharsalus too Pompey had positioned archers on his left wing to attack Caesar's right flank. In his African campaigns, however, Caesar several times stationed slingers and archers amongst the front ranks of his battle line so that, just as Onasander recommends, they could fire directly at the enemy before withdrawing to allow the heavy infantry of the legions to engage.[117]

Unfortunately historians of the Empire generally fail to provide these kinds of details, though it is clear that both slingers and archers continued to fulfil an important military role. Tacitus regularly refers to operations involving both types of troops, but when describing pitched battles rarely indicates where in the line they were positioned or what their roles were. Both slingers and archers are illustrated on Trajan's Column, the archers depicted in their pointed helmets distinctive of Syrian or other eastern auxiliaries. They are illustrated shooting at the enemy from behind other auxiliary troops. Onasander criticizes this arrangement because the archers were unable to fire with such force, but it could produce a hail of missiles to discourage the enemy from engaging, or could start to break up his battle line by inflicting casualties from a distance. Arrian planned to position his mounted archers at the rear of his battle line against the Alani. Firing over the heads of the eight ranks of infantry in front of them, these archers added to the missile barrage intended to break up the Alan cavalry charge. Additional fire-power was provided on the flanks, in the from of foot archers and artillery on the rising ground. In pitched battle at Issus during the civil war in AD 194, both Roman armies deployed with their heavy infantry in the centre and their artillery and archers at the rear to fire over the front rankers. The army of Septimius Severus was the first to advance under the hail of missiles, using the cover of a *testudo* formation, which is exactly what Onasander recommends for this situation.[118]

Though we are told less about their role in the histories of the imperial period, we can learn something of the importance of missile troops through the number of auxiliary units of archers. Some 32 units of archers are attested in epigraphic evidence in this period, including several units that were mounted or part-mounted. As indicated above, Arrian's army for his campaign against the Alani included a significant number of archers, both mounted and foot. These were an important part of his army given the highly mobile nature of his enemy and the tactics he intended to use in battle. Dedicated units of slingers are not recorded during the imperial period, though there were clearly slingers serving in the Roman army. Hadrian's address to the army in Africa suggests that practice in the use of slings may have been a general requirement and indeed Vegetius states that all new recruits should be taught to use the sling, and those with the aptitude the bow. These light missile troops could be extremely vulnerable if caught in the open by enemy infantry or cavalry, though the treatise writers do not warn of the dangers of this. It was important therefore that they received support from other infantry or cavalry. That may be why Roman generals regularly placed archers in the front line of their battle lines or mixed in with the infantry where the legionaries would protect them from being isolated. At Pharsalus, cavalry protected Pompey's archers and slingers, but once Caesar's cavalry drove them off, the light troops were left exposed, to be massacred by Caesar's advancing infantry. Auxiliary infantry at Idistaviso in AD 16 had been stationed to provide protection for the Roman archers, whom the Cherusci attacked in their efforts to escape.[119]

Cavalry

As we have seen, the principal role of cavalry in pitched battle was to face the enemy cavalry on the wings, prevent or execute flank attacks, and to carry out pursuit. Other roles included supporting light missile troops. For the most part, cavalry tended to avoid direct confrontations with heavy infantry, for sound tactical reasons: cavalry are usually unable to break such formations, hence the line of infantry proposed by Arrian against the Alan cavalry. The Nervian infantry in Gaul twice repulsed Caesar's cavalry, and his picked infantry at Pharsalus seem to have had little difficulty in forcing Pompey's cavalry to flee. We might question the skill and tenacity of the cavalry in both cases, and although Frontinus claims that Pompey's at Pharsalus was very skilled, this may be an exaggeration.

He records that Caesar ordered his men to aim for the faces and eyes of the cavalrymen to encourage them to flee. Once the cavalry had been driven off, it left the flanks of the infantry unprotected and vulnerable to attack by infantry, especially missile troops. It is in this respect that Caesar's alterations to his line at Pharsalus proved decisive, as his overloaded right wing with its fourth line put to flight the cavalry, then tore into the unprotected slingers and archers, before surrounding the Pompeians and attacking the rear of Pompey's left wing which had up until then still been resisting.[120]

During the late Republic, cavalry units frequently went into battle supported by light infantry. Several of Rome's enemies did the same, including Numidians and Germans, making their cavalry more effective against enemy cavalry. Since Caesar's cavalry was primarily Gallic and German it is perhaps not surprising to find that he too interspersed his cavalry with infantry, either light-armed auxiliaries or *antesignani*, legionaries used to fighting in the front rank of battle, but lightly armed for special duties. This infantry, small in number and lightly armed, could not have hoped to deter or hold a cavalry charge by the enemy in the way that Arrian proposed his legions would against the heavy cavalry of the Alans. They must then have acted in some other way to assist the cavalry, possibly by providing fire-power. Livy's explanation of the introduction of the *velites* to the legion at Capua in 211 BC may help to elucidate this. At Capua, the Roman cavalry experienced considerable difficulties in engagements with the Campanian cavalry, until they were 'stiffened' by the introduction of light-armed legionaries who acted in concert with the cavalry and provided them with firepower in the form of short lances:

> From all the legions young men were selected who were very active, lightly built and quick; they were given shields shorter than the cavalry ones, and seven javelins four feet long and with an iron head like the *velites'* spears. Each cavalryman took one of these men on his horse and taught him to ride behind him and dismount quickly when a signal was given. Once daily practice showed that this could be done smartly, they advanced on the plain between the camp and city walls against the formations of Campanian cavalry. When they got within weapons range the signal was given and the light-armed infantry dismounted. A line of infantry appeared suddenly from what had been cavalry and charged the enemy cavalry, throwing their javelins with great force. Very many of these were thrown at both horses and men and many of them were wounded; but greater fear was caused by these strange and unexpected tactics. The cavalry fell upon the shocked enemy and carried the pursuit and massacre up to the city gates. From that point on the Romans had superiority in cavalry as well, and the practice of having *velites* within the legions was established. It is recorded that the centurion Quintus Navius first had the idea of combining infantry and cavalry, and he was honoured by his commander for it.
>
> Livy 26.4.4-10

Although the mixture of cavalry and infantry at Capua and amongst Republican auxiliary cavalry units may have provided the inspiration for the part-mounted cohorts of the imperial period, the Republican units had a much higher proportion of infantry to cavalry, having more or less equal numbers of both. Because of the lack of detailed accounts of battles in the early Empire, it can be hard to determine how these mixed units actually fought in pitched battle, but the numbers involved, in theory 480 infantry to 120 cavalry, suggest that they were not fighting in the same way as the units at Capua. Arrian's battle orders indicate that the cavalry and infantry of his mixed units, or *cohortes equitatae*, operated entirely separately, the infantry with the rest of the auxiliary infantry, and the cavalry with the remainder of the cavalry on the wings. The infantry and cavalry of these mixed units were also separated when on the march, and the different exercises undertaken by the cavalry and infantry of the mixed units for Hadrian on his tour in Africa in AD 128 suggest that it was normal for them to operate independently. The existence of these mixed units in the imperial period was probably to allow tactical flexibility, particularly when such units were posted to garrison a fort or when small armies were required to deal with low level threats locally.[121]

Reserves

We have seen how the basic dispositions of the legion on the battlefield provided automatically a group of reserves for the Roman general. The treatise writers stressed the importance of such reserves to conduct flank and rear attacks and to deal with any threats to their own battle line. Use of such reserves, particularly the timing of their use, had to be judicious because once they were committed, there was very little a general could do to influence the outcome of a battle. At Munda, Caesar's veteran Tenth legion on the right wing was so successful that the Pompeians drew reserves from their own right to shore up their left wing. The now weakened left wing could not resist Caesar's infantry and cavalry and, since the Pompeians had no more reserves to throw in, they were forced to retreat. The treatises do not cover such topics in sufficient detail as to indicate exactly when reserves should be sent in. There were too many variables and possibilities for them to make any practicable suggestions: it is for the general or his officers to decide. Against Ariovistus, it was the cavalry commander Crassus who ordered in the reserves, because Caesar was fighting with the infantry. At Mons Graupius, however, Agricola did not actually enter battle himself and directed events from the rear. The entire legionary force was available to the general as a reserve force, but he did not need to engage them, success being brought about by the auxiliaries alone. He did, however, also have four cavalry *alae* for use in an emergency, and he threw these in to prevent a flank attack. The confusion of battle may have made the sending in of reserves a difficult task to judge, but in the case of both Crassus and Agricola, the actions and timing of ordering in the reserves proved decisive.[122]

Flexible Forces: variations and changes in the disposition of armies

Very few changes in the basic field dispositions are recorded from Polybius' descriptions of battles in the Second Punic War to the fourth century AD. There is a general agreement between the writers of military treatises and the accounts of battles in histories concerning the deployment and role of the different types of troops in pitched battle. Throughout the period, the image presented is that the legions usually carried the brunt of the fighting, flanked by allied or auxiliary infantry and with cavalry on the wings. However, two changes in the dispositions of the imperial period have attracted interest: the use in some battles of auxiliaries as the main striking force instead of the legions, and the use of so-called phalanx tactics. The reasons for the former are disputed, as is the evidence for the latter.

Auxiliaries as acies

In some pitched battles of the early imperial period, the auxiliary infantry units took the role of comprising the main part of the battle line which was usually held by the legionaries, whilst the latter were held in reserve. This method of disposition did not take over exclusively from the other arrangement already described, of legions holding the centre flanked by the auxiliaries: for example, both Suetonius' dispositions against Boudicca in AD 60 and Corbulo's proposed formation against Tiridates in AD 58 were of this type. This different method of deployment, however, using the auxiliaries as the principal striking force in pitched battle, can be seen at Idistaviso in AD 16, in a battle against the Frisii in AD 29, in Cerialis' defeat of the Batavians in AD 70, and most famously at Mons Graupius in the mid 80s. Tacitus' claim that Agricola used the auxiliaries in this way at Mons Graupius in order to gain a victory without the loss of any Roman blood has been accepted by many historians and has encouraged them to argue that the auxiliary troops in general were low-grade, expendable foreign troops. This, however, is not the case at all; the auxiliary infantry was a vital part of the army of the imperial period and provided reliable and versatile troops, as can be seen from the very use of them in the front line of battle.[123]

As we have seen, the military writers were aware that a number of factors dictated the deployment of a battle line, including the types of troops involved on both sides and the nature of the terrain. According to some ancient writers, the legions of the Republic were armed and trained specifically for the set-piece battle and had difficulty operating against lightly-armed enemies, particularly on rocky or uneven ground. Polybius and Livy both claim this when they describe engagements between legionaries and light-armed enemies, and try to explain the difficulties the Romans encountered. This theme is repeated by Plutarch when he notes the difficulties Metellus' legionaries faced against Sertorius' light-armed Spanish troops:

> Because of the agility and light arms of his Spanish troops, Sertorius was always able to adapt to changing circumstances, but Metellus was familiar with heavily armed troops and pitched battles fought by heavy infantry in a set formation. These troops were excellently trained for repelling and putting to flight an enemy in hand to hand fighting, but were unable to keep up with the incessant advances and withdrawals of light-armed men who moved like the wind.
>
> Plutarch, *Sertorius* 12

Tacitus too contrasts the heavy Roman legions with the more mobile and lightly-armed Cherusci in AD 16, though this engagement took place on marshy ground rather than uneven. Although to a certain extent this is a recurrent literary theme or *topos*, Roman legions clearly did experience difficulties in such terrain and against such light-armed and mobile enemies. The introduction of cohorts to the legion has been linked with Roman operations in Spain and the difficulties there of terrain and type of enemy. This may have allowed greater flexibility in the way legions and legionaries operated, but they continued to experience more problems when fighting on difficult or unfavourable terrain than some of their lighter armed enemies.

All the battles mentioned above, in which the auxiliary infantry replaced the legions as the principal striking force, were fought on terrain unfavourable for the Romans. At Idistaviso and Mons Graupius, the enemy was deployed on higher ground requiring the Romans to fight an uphill battle; the battles fought against the Frisii and Batavians were on very wet or boggy terrain. In the latter case the Batavian leader Civilis had deliberately flooded the battlefield in the knowledge that his troops, skilled in this type of fighting, would have had a considerable advantage over the Romans. Generals were advised by the textbooks to make best use of their troops for the nature of the terrain in which they were operating and the equipment of the auxiliary infantry allowed them to operate more effectively in certain types of terrain than the legions. This variant in the organization of the battle line should be seen as illustrating not a desire to preserve the legions at the expense of non-citizen troops, but the versatility of the auxiliary units and their strength as a fighting force.[124]

Legion as phalanx

Arrian's *ektaxis* is the most detailed description of a Roman army on campaign and its tactics since the accounts of Caesar and others in the late Republic. His proposed deployments for battle have given rise to considerable discussion and encouraged some historians to make the rather surprising suggestion that in the second century AD the Roman legion deployed in a way very similar to the Greek phalanx rather than using the formation of cohorts explained above. Arrian proposed to deploy his legions not in the 4-3-3 cohort formation usually found in the late Republic and early Empire, but in an infantry formation eight ranks deep. The ranks were packed much more closely together than usual, the intervals between the ranks being 1½ feet in contrast to the three feet interval suggested by Vegetius. The front four ranks were armed with one type of spear (the κόντός) and the front rank aimed these at the bellies of the enemy horses. The next three ranks, whose spears would have extended beyond the front rank as in a phalanx, also used their weapons for thrusting whilst the rear four ranks hurled their *pila* (or λόγχαί) over the heads of those in front of them. Archers were stationed at the rear and also fired over the heads of the legionaries; cavalry and auxiliary infantry held the wings, and archers and artillery, stationed on the hills on which Arrian proposed to fix his line, provided more firepower.

In many respects, Arrian's proposed battle line is not particularly unorthodox. The positioning of the legions, with auxiliaries and cavalry on the wings is close enough to the 'standard' deployment of armies in the late Republic and early Empire, and the firepower behind the heavy infantry and on rising ground is both recommended by treatises and seen in a number of pitched battles. It is the deployment of the legions, their weapons and Arrian's proposed legionary tactics that appear new, and it is this that has led historians to suggest that the legion deployed as a phalanx in this period. We should be cautious, though, in accepting that this represents a radical change in the way legions were armed and functioned in battle. Certainly changes in armaments and fighting practices did occur, but there is no evidence for an empire-wide or sudden reform in the second century AD. Arrian's equipment and deployments may simply represent a one-off 'stratagem' to deal with a particular tactical situation, or a regional variation because of the nature of the enemy in that area.

Although the different armaments of Arrian's legionaries marks a change from the usual uniform equipment of the cohortal legion, legionaries had in the past been equipped differently, in the manipular legion, and Caesar's *antesignani*, the picked legionaries he detailed for the duties of light-armed troops may also have used different equipment. There was clearly a certain amount of experimentation in the design of weapons in the Roman army, particularly with *pila* and spears. Marius and Caesar both made alterations to the *pilum*, Sallustius Lucullus, governor of Britain in the late first century AD named a new type of spear after himself (the unlikely reason Suetonius gives for his execution by Domitian). Archaeological evidence indicates a considerable variation in the design and size of *pila* and spears. Such changes do not necessarily indicate a major change in fighting technique, and those of Marius and Caesar were little more than refinements in the design of the *pilum* necessitating no change in the way they were used. There is some evidence to suggest a certain amount of specialization of arms within the legion by the third century AD, such as the light spear-men or *lancearii* from Legion II Parthica based in Apamea in Syria, but this is some time after Arrian's campaign against the Alans. In Arrian's case, there is no indication that he was not using the usual *pila* and spears of the legions and auxiliary units, or that his soldiers were always armed in this manner. Just as Agricola and other generals of the first century AD used auxiliaries to solve a particular tactical problem related to topography, so perhaps Arrian was equipping and deploying his legions to tackle the problem of facing heavily armed cavalry. Arrian's aim in his battle plan was to prevent the Alan cavalry from breaking his own battle line whilst forcing them to turn and flee under pressure from the hail of missiles. Because cavalry will not generally charge into a solid mass, particularly one bristling with spear points as Arrian's was, this kind of formation was the most effective one against an enemy comprising heavy cavalry.[125]

It was not uncommon for Roman armies to develop and refine their tactics and fighting techniques to face different enemies. We have already seen how the concept of adopting the enemy's weapons and adapting them for their own use was a popular theme in histories of the Roman Republic and later, but there are other examples of variations made for a particular battle or campaign.

In his critique of the Roman tactics against the Insubrian Gauls in 223 BC, Polybius puts the Roman victory down to instructions on tactics and hand-to-hand fighting given by the military tribunes. The tribunes were anxious that their men should not be face-to-face with the first onslaught of the Gauls when the latter were eager for battle, so devised tactics that would dampen both the enemy's eagerness and the effectiveness of their swords. Polybius claims that the Gallic swords were of such poor quality that they blunted very quickly. He is clearly exaggerating the uselessness of the Gallic swords, but the size of them prevented the Gauls from fighting effectively at very close quarters, and according to Polybius this is what the tribunes exploited:

> The Romans are thought to have shown great skill in this battle because of the instructions from the military tribunes to their men concerning general combat and hand-to-hand fighting. They had noticed in earlier battles that the Gauls in general were at their most dangerous in the first attack when they were still fresh, and that because of the way their swords are made, as mentioned above, only the first strike has any effect. After this they immediately take the shape of a strigil, the blades being so bent both lengthways and sideways that unless the men are given some opportunity to place them on the ground and straighten them with the foot, the second blow is completely useless. The tribunes therefore distributed to those in the front ranks the spears of the *triarii* who are stationed in the rear, and ordered them to use their swords only when the spears were done with. Then they drew up opposite the Celts and engaged them. Once the Gauls had rendered their swords useless by slashing at the spears, the Romans closed with them and rendered them helpless by denying them the room to slash with their swords; this stroke is unique to the Gauls, and their only one, because their swords have no points. The Romans, on the other hand, did not use slashing moves, but instead used their swords in a straight thrusting motion, using the sharp points which were very effective. Striking one blow after another at the chests and faces of the enemy, the Romans killed most of them.
>
> Polybius 2.33.1-6

Interestingly, Vegetius notes that it was again the tribunes who were responsible for training the soldiers at the start of a campaign. He may have got this information from one of his Republican sources, perhaps Cato the Elder's treatise, which was roughly contemporary with Polybius and one of Vegetius' sources of information. Trajan himself undertook the training of his soldiers on their long march to Parthia, no doubt to prepare them for the different styles of fighting that they would be facing. Just as units in a particular area picked up local customs, so they picked up and adapted to local fighting practices, something that is apparent from the accounts of Caesar's civil war campaigns. We have already seen how his line of march encountered difficulties in Africa because of the harrying tactics Scipio employed. Caesar was forced to halt the campaign temporarily to re-train his men so they could cope with the demands of a different fighting style.[126] Following an unsuccessful engagement at Ilerda, Caesar points to another 'unorthodoxy', in the fighting techniques of the Pompeians in Spain:

> Their method of fighting was to run forward with a great charge and boldly capture a position, not worrying about preserving their ranks, but fighting in small, scattered groups. If they were being pressed, they did not think it shameful to retreat and yield the ground. They had grown used to fighting with the Lusitanians and other barbarians in a kind of barbarian battle style; for soldiers who have spent long periods in particular places are often influenced by the customs of those regions.
>
> Caesar, *B.Civ.* 1.44

Arrian's *ektaxis* is the only detailed description of a battle line between Tacitus' account of Mons Graupius in c.AD 84 and Dio's of the civil war battle at Issus in AD 194 in which legionaries deployed in a fairly orthodox manner with missile troops stationed behind them; we should not place too much emphasis on one isolated example of variation. Given the size of the Roman empire, the variations in topography and climate, and the different enemies with different fighting styles that Roman armies encountered, it would be far more surprising if there were no variations in the way Roman armies fought; Arrian's deployments should be seen in this context. In accordance with the advice of the textbooks he was making best use of his troops and the terrain to face a particular type of enemy. Such tactics would have been unnecessary or ineffective against other enemies in other provinces, particularly those with strength in infantry rather than cavalry. We should be cautious then of interpreting Arrian's dispositions as representing a permanent, or empire-wide, change in military practices. What his tactics illustrate most clearly is the continuing flexibility of Roman armies and their willingness and ability to adapt to different situations.

Retreat and pursuit

> When the enemy has been repulsed, if it is clear that they are fleeing, the infantry will open up their ranks for the cavalry to advance, though not all the companies, but only half of them. The other half will follow those in front, but more deliberately and in ranks, so that if the rout becomes a full one, those in the front of the pursuit can be relieved by fresh cavalry, and if the enemy wheel about they can be attacked as they turn. At the same time the Armenian archers should advance and fire, to prevent the fleeing enemy from turning back again, and the light-armed spearmen should follow at the run. The infantry formation will not remain on the battlefield but will advance at a quick march so that if stiff resistance from the enemy should be encountered, they can again become a protective screen in front of the cavalry.
>
> Arrian, *ektaxis* 27-29

The line of battle was designed to repulse or, if possible, break the enemy, but it was the cavalry and light infantry that carried out the pursuit of a fleeing enemy. The treatise writers stress the importance of keeping the battle line of heavy infantry in formation during both retreat and pursuit. Maintaining ranks was the best way to keep casualties to a minimum in both battle and its aftermath, as Marius' recruits had learned during their training in Africa. An army engaged in pursuit was advised to stay in formation in case the enemy turned and renewed the fight. An army that had abandoned the close formation of its battle line and was in full flight made easy pickings for the pursuing cavalry and archers, and this was the point at which the majority of casualties would occur. Vegetius suggests that the defeated enemy should not be completely surrounded but allowed to flee so they would be less likely to turn and renew the fight, and Frontinus devotes a section of his *Stratagems* to this subject. He illustrates how if the enemy was totally surrounded he was more likely to fight to the death, a course that could result in heavier casualties to the eventual victor.[127]

46 Cavalry pursuing the defeated Dacians and their king Decebalus, Trajan's Column. Courtesy JCN Coulston

Roman armies had various formations available in case of the need to retreat, including the *testudo*, and a defensive circular formation, the *orbis*. A pursuing enemy would be more likely to concentrate on easier more scattered targets, and so these formations were particularly useful if the battle line had been broken up and groups of soldiers were attempting to retreat to safety. Domitius Calvinus' Legion XXXVI escaped from the defeat by Pharnaces in 47 BC by forming a circle and withdrawing to higher ground where the pursuers were reluctant to follow. The other 'Roman' forces, mainly local levies and the 'legions' of Deiotarus, suffered severe losses because they lacked the discipline and training to undertake such complex manoeuvres during the stress of defeat in battle. However, the principal reason for the construction of a marching camp by Roman armies near to the proposed battlefield was to provide a defensible position to which defeated troops could retreat. We have seen how in civil war encounters between two Roman armies, the proximity of such camps to the battlefield might even lead to the avoidance of battle simply because, with a refuge so near there was no chance of achieving a decisive victory through the mass slaughter of a defeated army in full retreat.[128]

Pursuit was carried out by the cavalry, as the textbooks state, sometimes accompanied by the light infantry, and historical sources frequently refer to the heavy casualties caused by pursuing cavalry **(46)**. Caesar's albeit small-scale successes against the British in his first expedition were severely limited by his absence of cavalry. He was unable to pursue the Britons very far when they fled from two separate engagements with his infantry, and so could not score any decisive victory over them. Arrian proposed that his pursuit of the Alans should be carried out by the cavalry and light infantry, his 'phalanx' or battle line of legionaries was to remain in formation and advance slowly behind the pursuers in case the enemy should turn and renew the attack, just as Onasander recommends. Pursuit might continue for a considerable time: at Mons Graupius the pursuit did not stop until nightfall whilst after Strasbourg in AD 358 Julian's men pursued the Germans to the river Rhine. After Germanicus' success at Idistaviso the pursuit continued for some ten miles until it too was stopped by a river and by nightfall:

> The rest were massacred. Many tried to swim the Weser but were hit by javelins or carried away by the current, until finally they were overcome by the mass of those fleeing and collapse of the river banks. Some cowards tried to flee by climbing trees and as they hid among the branches archers picked them off for fun; others were brought down by felling the trees.
> It was a great victory, at little cost to us. The slaughter of the enemy continued from the fifth hour (about midday) till nightfall, their bodies and weapons scattered for ten miles around.
>
> Tacitus, *Annals* 2.17-18

Though the comprehensive defeat and demoralization of the enemy was the principal reason for carrying out a thorough pursuit and slaughter of those fleeing, it was not the only one. As with the massacre of both defenders and civilians in towns taken by storm, it gave the soldiers the opportunity to sate their blood-lust and increase the amount of booty taken by the successful army, though the majority of the spoils were likely to have been obtained through the capture of the enemy encampment. The numbers of casualties in a pitched battle provided the simplest means by which the size of victory or defeat could be judged. The greater the slaughter, the greater the victory, and in the Republic the size of the victory might help to determine whether or not a general was awarded a triumph, the right to process through Rome in celebration of the victory. In the second century BC, legislation was passed requiring at least 5000 enemy dead before a commander might be considered for a triumph, though an application to triumph might be refused if the Romans too suffered heavy losses: Appius Claudius Pulcher was refused a triumph by the senate in 143 BC following his defeat of the Gallic Salassi because although some 5000 of the enemy were slain, so was an equal number of Romans. In the Empire, senators could no longer celebrate a triumph, that honour becoming the prerogative of members of the imperial family. Instead, consular governors might be awarded triumphal ornaments, but even these might be awarded for actions involving little or no military activity. Thus the casualty figures on both sides must have been one of the only available means of publicizing the importance and extent of victory in pitched battle for a successful general of the imperial period.[129]

Casualty figures reported by ancient historians are notoriously inaccurate, as the historians themselves acknowledged. Livy frequently criticizes the first century BC historian Valerius Antias for exaggerating the numbers of enemy dead, though he still quotes his figures. There were reasons for being less than accurate with such figures, to stress the magnitude of the victory and the small numbers of Roman casualties, for both political and literary reasons. Strabo, for example, claims that when Q. Fabius Maximus defeated an army of Gauls in 121 BC, 200,000 were killed. Even Appian's figure, at 120,000 the most conservative estimate, seems pretty fantastic, especially given his figure of only 15 for the Roman losses. However, the casualty figures of victors and losers were frequently very disproportionate in Roman warfare: at Thapsus, the Pompeians lost 5,000 to Caesar's 50; at Munda 30,000 to Caesar's c.1000; against Boudicca, 80,000 Britons to 400 Romans, and at Mons Graupius, 10,000 to 360 Romans. Even if we are very sceptical about the accuracy of all these figures, both winners and losers, and we must be, the proportions indicate the extent of the victories and the huge damage that could be inflicted on an enemy army once it had turned and was in flight.[130]

The general

> When some said that Scipio Africanus had too little aggression, he is reported to have replied, 'my mother bore me to be a general, not a fighter.'
>
> Frontinus, *Stratagems* 4.7.4

> There was a difference between the soldiers' duties and the general's: an eagerness to fight was appropriate for soldiers, but generals were more often of value by being prepared, deliberating and showing caution rather than recklessness. He (Antonius Primus) had formerly contributed to victory through courage and arms, but he would continue to do so through planning and sense, the skills characteristic of a general.
>
> Tacitus, *Histories* 3.20

As he was writing a treatise on generalship, Onasander of course gives prominence to the position of the general and his role in pitched battle: 'when fighting, the general should show caution rather than boldness, or he should avoid completely hand-to-hand combat with the enemy'. There are several reasons for this strong advice that the general should avoid becoming too closely involved in the actual fighting. Although Onasander recognized that the presence of the general could encourage his men in battle, the blow to their morale should he be killed or seriously injured could be such that the potential disadvantage of his presence in the thick of battle far outweighed the advantages. In addition, if the general were fighting in the battle line, he would be unable to direct his troops, react to emergencies and send in reserves where necessary. Onasander uses the analogy of a ship's captain leaving the helm untended to do the job of an ordinary sailor. The general was expected to ride by and encourage his men by his presence and move his troops around the field when necessary. In the event of defeat, however, the general was expected to fall with his men, as Aemilius Paullus did at Cannae, Crassus at Carrhae and Varus in Germany in AD 9.[131]

In the hoplite phalanx of the classical Greek city-states, generals were expected to lead from the front and because of the nature of hoplite warfare, they not surprisingly tended to suffer high casualty rates whether on the victorious or defeated side. Theoreticians of the fourth century BC suggested that instead the general should direct the action from the rear of the phalanx; this was partly to ensure his survival, but also to direct the increasingly complex tactics of pitched battle in the fourth century and later. Such advice was not necessarily followed, and in later periods kings and generals, particularly Alexander and Pyrrhus, were noted for their willingness to lead their men into battle at the head of the charge. Roman generals too fought in the front ranks, as Aemilius Paullus did at Cannae, first with the cavalry on the right, and then in the centre of the legionary line, which looked at the time as if it would be victorious. There seems to have been a tendency amongst Roman generals of this period and later to position themselves at a point in the line that was likely to be decisive; or they directed the action from further back and might join a particular section if it was in difficulty or was about to achieve a breakthrough.[132] Frontinus reports how Sulla took the field to shame and encourage his men into renewing the fight:

> When his legions were yielding to the army of Mithridates led by Archelaus, Sulla advanced with drawn sword into the front line and, addressing his soldiers, said that if anyone asked them where they had left their general, they should reply, 'Fighting in Boeotia'. Shamed by these words, they all followed him.
>
> Frontinus, *Stratagems* 2.8.12

Frontinus reports Caesar doing something similar at Munda, and in some versions of these stories, the general sends his horse away, signifying that he did not intend to flee but would fight and win or die with them. Ammianus claims that Julian did the same at Strasbourg when he chased after his fleeing cavalry and encouraged them to renew the fight. Ammianus makes a direct comparison with Sulla's actions at Orchomenos, 'allowing for some differences'. Actually, Julian does not seem to have taken any serious part in the battle itself, and was accompanied by a mounted bodyguard of some 200 when he checked his cavalry's flight, so the historian's comparison is very loose. Compared with the Republic when actions like Sulla's and Caesar's are fairly common, in the imperial period there are far fewer reports of generals fighting in the front ranks with their men.

This may be because of the nature of our sources. Accounts of pitched battles by historians such as Tacitus and Dio tend not to be as detailed as earlier writers and may simply have not included details of where and when, if at all, the general became involved in any hand-to-hand combat. They continue to note the heroic conduct of soldiers of lower ranks, such as the centurion Julianus who charged the Jews alone at Jerusalem when the Romans were giving way, or the two unnamed Flavian soldiers at the Second Battle of Cremona in AD 69 who lost their lives putting an enemy catapult out of action. It is more likely that generals of the imperial period avoided this kind of behaviour. They may have felt there was less need to risk themselves in this way. Such actions were only necessary if a part of the battle line came under considerable pressure. Although the battle accounts heighten the drama by stressing the strength of the enemies Roman generals were facing, there is little evidence of Roman armies being very hard pressed in pitched battles in the early Empire. Lower ranks continued to be rewarded for conspicuous bravery by decorations, promotion and bonuses, but for the elite there were no opportunities to triumph and little to gain politically from such actions. The limited rewards to be gained through successful military service in the Empire may have discouraged the kinds of actions that in the Republic might have brought death, or hugely increased prestige in the case of success.[133]

If the general fought as a foot soldier, other officers would have had to ensure reserves were sent in when necessary. Caesar's presence with his right wing against Ariovistus greatly encouraged his legionaries there, but he may have been lucky that his cavalry commander, Crassus, noticed that the left was in difficulties. He ordered the third line of the *triplex acies* in support of the left wing, which helped ensure a total victory. The loss of the general in the fighting could, as Onasander warns, thoroughly demoralize an army, and even if he was just wounded, the news could influence the outcome of battle: when the consul Pansa was wounded in action against Antony, his army turned and fled. Frontinus includes comparable examples. In a similar vein, both he and Onasander recommend spreading rumours in the middle of battle. Even if they were untrue, such as falsely claiming that the enemy leader was dead or wounded, they could increase the morale of the army and demoralize the enemy. Injuries to generals did not always have negative effects on armies, however: when Metellus was wounded fighting Sertorius in Spain, his soldiers were ashamed and once the injured general had withdrawn from battle, they fought much harder than previously.[134]

After the battle

For Vegetius, once the defeated enemy has fled the battlefield pursued by the cavalry, the battle is over, and he provides no further information on conducting a field campaign. Onasander, however, covers a number of important matters in the aftermath of battle; namely rewarding the brave, control of plunder and burial of the fallen. Rewarding brave conduct, and punishing those who showed cowardice, acted as an inducement to fight bravely in battle and not to give way. Units that did give way in battle might be ordered to camp outside the security of the fortifications and distanced from the 'community' of the army, and receive barley rations instead of corn. In extreme cases they might suffer decimation, in which every tenth man, chosen by lot, was executed. Rewards, as Onasander suggests, were granted according to rank, with promotions, monetary rewards and decorations given by the general or, in the case of the Empire, in the name of the emperor. Plundering the enemy camp and dead was also a reward for the soldiers. The 'official' plunder was at the disposal in the Republic of the general, who was nonetheless expected to pass on at least a share to his men, and might be prosecuted for failure to do so. In the Empire, however, booty seems to have gone to the imperial treasury, but there must have been plenty of opportunity in both periods for 'unofficial' plunder to find its way into the soldiers' possessions.[135]

Such actions and opportunities may have helped to maintain or increase the morale and confidence of an army, and the burial of the fallen would also have been an important factor in army morale in the aftermath of battle. Onasander suggested that soldiers should go into battle in the knowledge that if they were killed they would receive a decent burial; they would then be more willing to fight and, if necessary, die. The customs and traditions of Classical Greek warfare generally allowed both sides to retrieve and bury their dead after a battle, though usually the defeated would have been despoiled by the victors, before they could be retrieved. As the scale of warfare and pitched battles increased in the Hellenistic and Roman periods, however, this might not always have been possible, particularly after a very serious defeat, such as Cannae or Cynoscephalae. Plutarch indicates that the bodies of the Germans killed by Marius at Aquae Sextiae in 102 BC were left where they had fallen, for he says that the fields where the dead bodies had rotted away over the winter were very fertile, providing bumper harvests in the following years! Unfortunately historians of Roman campaigns tend to provide far fewer details of these activities than their Greek predecessors. Livy occasionally, but not always, comments on the collection of their dead by the victorious side, and their burial, usually the day after battle, but few other details are offered. Frontinus cites one example where a Roman general buried many of his men by night immediately after the battle so that when the Spanish collected their dead the following day, it appeared that far more of their own had been killed than the Romans. Demoralized by this apparent defeat, they came to terms. Particular efforts may have been made to secure and bury generals and senior officers who had been killed, or lower ranks who had shown conspicuous gallantry. The defeated Othonians found and buried one of their legionary commanders who had been killed fighting at Cremona; and Caesar ensured that the centurion Crastinus, who had died leading his men at Pharsalus, was found and buried with his posthumously awarded decorations, in a separate tomb near to the mass grave of the other casualties. Livy claims that Hannibal had the body of Aemilius Paullus found and buried after Cannae, and had attempted the same with Flaminius' body after Trasimene, but it could not be found.[136]

Few memorials to the Roman dead are known compared with those set up by Classical and Hellenistic Greek city-states **(47)**. Germanicus had the dead of the Varian disaster buried when his own army came across the site six years later, and had a turf funeral-mound constructed. The dead of Trajan's Dacian campaigns were commemorated by a monument at Adamklissi in Romania, where inscriptions listing the casualties were also set up. If campaigns were being conducted in Italy, some of the dead may have been returned to Rome for burial: early in the Social War when Rome suffered heavy casualties in an ambush, the bodies were brought back to Rome, including that of the consul Rutilius. However, the numbers involved and reactions of the population caused the Senate to decree that casualties of the war should be buried where they fell and not brought back to Rome, so that the people would not be deterred from entering military service themselves. Proper burial of the dead usually served to increase the morale of the army, but as with this example, it could backfire. Livy explains how Philip V of Macedon gave a public funeral to some fallen cavalrymen, assuming that, as Onasander suggests, this would increase his popularity with his men and make them more ready to face danger on his behalf. But when his soldiers, used to the comparatively small puncture wounds caused by spears and arrows, saw the appalling injuries caused by the Romans with their Spanish swords, they were even more reluctant to fight.[137]

The time after a battle spent dealing with the dead also allowed for the recovery and treatment of the wounded. Here the victor left in control of the battlefield would have had the advantage in being able to reach his own injured first, and despatching the enemy

47 *Victory monument of heaped up arms and equipment, Trajan's Column.* Courtesy JCN Coulston

wounded, as Hannibal's soldiers did to the Roman injured at Cannae. After defeating the Helvetii, Caesar spent three days seeing to his wounded and burying his dead, delaying his pursuit of the large number of Helvetii who had escaped the battle. Trajan is supposed to have torn up his cloak to provide bandages for his injured soldiers in the Dacian War, further cementing the relationship between emperor and his troops.[138]

Conclusions: the limitations of theory

> When the storm of war is at hand, destroying, fluctuating and bringing different situations, the sight of the events as they happen demands inventiveness appropriate to the occasion, which the necessity of fortune suggests rather than experienced memories.
>
> Onasander 32

> The battle line was dictated more by the nature of the terrain, the slope of the hill and the demands of the immediate situation than by military theory and tactics.
>
> Caesar, *B.G.* 2.22

On the whole the advice provided by the theoreticians on the conduct of pitched battles is sound, and well illustrated by accounts of battles in histories. The concerns expressed in the handbooks, particularly about the nature of the terrain and the dangers of being outflanked, are also of significance in descriptions of the topography of the battlefield and dispositions for battle. There are inevitably variants, and occasions when what the treatises suggest does not reflect actual practices, but as both Onasander and Caesar point out, it is simply not possible to reduce everything to a set of rules and regulations. Whatever the theory might propound, sometimes it was simply necessary to make the best of an unexpected situation and make up a plan there and then: textbooks cannot cover everything, and this is something that Onasander accepts. Vegetius is very rigid in his advice, supplying his recommended 'standard' battle line, plus a limited number of alternatives depending on topographical circumstances and the nature of the troops on both sides. Onasander is much more flexible, content to provide general advice only and leave the commander to make many of the decisions, particularly concerning dispositions.

Vegetius' approach is typical of the cautious nature of his work, concentrating as it does on defence. All the military writers advise care when making the decision to engage in pitched battle, and Onasander suggests avoiding battle until a suitable place is found, but only Vegetius recommends avoiding battle completely and only fighting when absolutely necessary or when the outcome was in no doubt. This approach is more in line with the cautious defence of the later Empire than the more risky offensive strategies of earlier periods, and the tone of the treatises therefore reflects, to a certain extent, the periods in which they were written.

Roman battle tactics tended to be very conservative; the fundamentals of choice of terrain, dispositions and engaging the enemy changed little over the period under study. Variations are clearly visible in different periods and different geographic areas of the empire as Rome faced different types of enemy and had to adapt to the terrain. These variations, such as the tactics used against the Insubrian Gauls, at Mons Graupius, and proposed by Arrian, illustrate well not only the flexibility of the Roman army, but also the tactical awareness and imagination of Roman generals and officers. But these variations are what Frontinus would have called 'stratagems', plans devised by the commander (or officers) to resolve a particular situation. Such stratagems might be of value on other occasions too, or inspire the general to devise his own solution but they should not be interpreted as representing a permanent or empire-wide change in organization, procedure or tactics.[139]

5 Violent Confrontations: siege warfare

Introduction

> Siege warfare requires courage on the part of the soldiers, knowledge of military science and availability of military engines.
>
> Onasander 40

Siege warfare in the Roman world was a brutal and bloody affair, with very high stakes. Sieges could be decisive in the outcome of a war: the surrender of Alesia in 52 BC and capture of Jerusalem in AD 70 marked the end of serious resistance to Rome by Gallic and Jewish rebels respectively. This form of warfare threatened not just the lives of the soldiers, but the existence of entire communities. Siege warfare in antiquity was total war. Old men, women and children, even animals might be slaughtered indiscriminately in the chaos of a city being sacked. At Jerusalem Roman soldiers killed so many that they grew sick of the slaughter; the city itself was totally demolished, only three towers of the wall surviving. The 700 year old city of Carthage was captured by Scipio Aemilianus in 146 BC. Scipio might have wept at the sight of a city in flames, but it was still razed to the ground and a curse placed on anyone who settled there. Those inhabitants who had survived the massacre were homeless and without property.

The difficulties of siege warfare often turned them into violent contests of ingenuity as attackers and defenders sought to anticipate and counter each other's moves. The violence and drama of these competitions of skill and wit with their catapults, artillery, battering rams, siege towers, mines and cunning stratagems made them irresistible subject matter for historians and experts alike. They describe sieges at length, and historians sometimes break off the narrative to explain in detail the amazing engines and contraptions used. Ammianus halts his account of Julian's invasion of Persia once the emperor has crossed the Tigris, to describe the various types of siege engines that will appear in the following chapters as Julian attacks and captures Persian fortifications. Ammianus himself had experienced and survived the siege of a town, that of Amida on the upper Tigris in AD 359. He gives an exciting and extraordinarily vivid account of his flight into the city, escaping the Persian van by playing dead below the ramparts of Amida, then fleeing into the city under cover of artillery fire. He describes the blockade and assaults up until his

48 *A Roman style assault using the famous* testudo *formation, Trajan's Column.* Courtesy JCN Coulston

escape just as the city was falling into Persian hands. The historian Josephus was commander of the Jewish forces at Jotapata, which was besieged and taken by the Romans by storm in AD 67. Then Josephus, having seen Roman siege warfare from the defender's viewpoint, witnessed it from the other side, as a prisoner in the Roman camp at Jerusalem. Fascinated by the workings of the Roman army, Josephus' accounts of what he witnessed are highly detailed, often very graphic, and sometimes gratuitous, though not always trustworthy. Caesar too provides us with eye-witness accounts of many sieges during both his conquest of Gaul and the Civil War. Most famous are the sieges of Avaricum and Alesia, the one taken by assault, the other forced to surrender, which form the twin centrepieces of Book Seven of the Gallic War.

The capture of towns and strongholds in enemy territory was of key importance in Roman warfare and centres of occupation were usually one of the principal objectives of an invading force. Their value if captured was even greater, if they were sited in strategically important places, such as Dura Europos on the Euphrates, or Syracuse in Sicily. Occupation of Sicily was central to the control of the Western Mediterranean and Syracuse provided not only a safe harbour but also a stepping-stone between Africa and Italy. The city's alliance with Rome in the First Punic War had enabled Rome to conquer the island, but when it joined Carthage in the second war, southern Italy became vulnerable, and Syracuse became a means by which Carthage could have reinforced Hannibal in Italy. Rome sent Marcellus, arguably her best general, to re-capture Syracuse, whilst others kept Hannibal occupied in the Italian peninsula. Its capture in 212 BC provided the military and supply base from which Scipio could launch his successful invasion of Africa eight years later. In preparation for this expedition in 204 BC, Syracuse was stuffed full of military supplies and men, ready to launch a combined land and naval attack. New Carthage (Cartagena) had provided a similarly important supply base for the Carthaginians in Southern Spain. Its capture by Scipio in 210 BC by a daring surprise attack must have been as much a blow to the Carthaginian treasury and supplies as it was a windfall to the Romans. Indeed, when providing the detailed official list of plunder from the city, Livy notes that the military supplies, along with the gold and silver, were considerably more valuable to the Romans than the city itself. Scipio's success also encouraged a number of Spanish tribes to switch allegiance to Rome, shifting the emphasis of the war in Spain and bringing closer the Roman invasion of Africa.[140]

The wealth to be gained from siege warfare could clearly be considerable, and the Roman treasury would have benefited greatly from the capture of a wealthy city like New Carthage or Corinth. The soldiers too could be rewarded for their efforts by enriching themselves during the sack of a town; Josephus claims, somewhat dubiously, that the value of gold in Syria halved because the Roman soldiers had so much of it following the capture of Jerusalem. It was certainly good for morale to give soldiers the opportunity to pillage captured towns, and this would have increased the popularity of a general with his troops and cemented their loyalty. Whilst military and strategic considerations were usually the reasons for an attacking army to besiege a stronghold, the potential financial rewards of success may have been an added encouragement to both commander and soldiers, and would have helped to defray the expensive cost of siege warfare. Perhaps more importantly though, wealth captured meant wealth denied to the enemy and reduced his ability to make war. This could be very significant in the course of a war. Marius' capture of Capsa and other towns in Numidia on which Jugurtha had based his defensive strategy required the king to reverse his plans, which were clearly not working, and seek pitched battle with the Romans. But the loss of his treasury when the Romans captured the Numidian fortress on the river Muluccha cost Jugurtha his Gaetulian mercenary force which he was no longer able to pay, and led him to force his reluctant son-in-law Bocchus into an alliance. Bocchus subsequently betrayed Jugurtha to the Romans.[141]

Siege warfare could be the only way to attack an enemy who declined to accept a pitched battle, or who had been defeated in the field and taken refuge in a well-defended city. The Jewish strategy in the revolt of AD 67-70 was to avoid direct conflict with the Romans in pitched battle. With little training and poorly equipped, the Jews would almost certainly have been incapable of withstanding a pitched battle against the well-armed and trained Romans. Hit-and-run tactics had forced Cestius, the governor of Syria, to retreat ignominiously from Judaea after abandoning his siege of Jerusalem, and with the arrival of Vespasian, the Jews based their resistance on the defence of their cities. Such a strategy was a more effective one against the Romans than accepting pitched battle, but it was one that delayed rather than prevented the crushing of the revolt. It was rare, however, for a campaign to be conducted so extensively through siege warfare.

The character of siege warfare varied with the topography of Europe and the Mediterranean, and with the nature of the enemy. Besieging a well-defended city or fortress in the semi-desert climate of Judaea was a very different prospect to storming a hill-fort in Gaul in terms of both hardship and technique. Many city-states on the Mediterranean littoral had sophisticated defences developed in response to the advanced siege techniques of the Hellenistic period. Cities such as Syracuse, Carthage and Jerusalem contained different areas with independent defences or a citadel, making their capture a gradual and dangerous process. The idea of siege warfare and techniques of attack and defence were not new to these cities, again increasing the dangers of a siege. Stockpiling of food and military supplies in preparation for a siege, and internal water supplies also improved a city's chances of withstanding a siege. Likewise, the availability of food, water and other supplies influenced the effectiveness of an attacking or blockading army and therefore possibly the outcome of a siege. Titus' army at Jerusalem stripped the surrounding land of its timber to build siegeworks, and the need to collect it from further afield presumably caused some delay to the progress of the siege, but Caesar was forced to abandon his (albeit not very effective) blockade of Pompey at Dyrrachium because his own army was running out of food.[142]

In the areas of the Roman Empire that were not urbanized in the way of the Mediterranean city states, such as northern Europe, many of these problems did not exist, but there were others. The defences of hill-forts, what Caesar usually refers to as *oppida*, were generally not as sophisticated as those of cities, many had no internal water supply and since there was no tradition of the type of assault and blockade techniques familiar to the Mediterranean, such actions could come as a nasty surprise to the defenders. Early in Caesar's campaigns in Gaul, the Suessiones were so alarmed by the Roman siegeworks that were new to them that they surrendered their *oppidum* of Noviodunum in northeastern Gaul. Some hill-forts, such as those in Britain, were taken quickly and relatively easily through direct assault. Others presented more of an obstacle. Caesar describes the *murus Gallicus* or Gallic wall that protected a number of *oppida* in Gaul. The wall or rampart was constructed of a latticework of timber with a rubble filling and stone facing. The stone protected the construction from fire and the timber against battering rams. This type of fortification hampered the Roman efforts at Avaricum. Although the siegeworks at Avaricum were, by Caesar's account, on a very large scale, most others were nowhere near the scale of works at cities like Syracuse, Marseilles or Jerusalem. On the other hand, the relatively mobile wealth of Celtic tribes meant that the capture of an *oppidum* might not have such an effect on the enemy's ability to make war. In his campaign against the Veneti, for example, Caesar was regularly frustrated because whenever his troops were about to capture one of their coastal hill-forts, they would abandon it and, along with all their possessions, sail on to another one. Caesar abandoned this strategy when he realized that despite capturing several of the *oppida* he was getting nowhere and could not harm the Veneti, because he could hit neither them nor their resources.[143]

Precursors to Roman siege warfare

Most of the basic techniques of siege warfare that we are familiar with from the Roman period were being used in the Eastern Mediterranean and Middle East from as early as the ninth century BC, and probably from considerably earlier. There is good evidence dating from this period and later, showing a gradual evolution in the development of siege technology and equipment. Assyrian and other early sieges are described in literary records and are graphically illustrated in carved reliefs from royal palaces at Nimrud and elsewhere in Assyria, dating from the ninth century BC and later. The reliefs show a range of equipment and techniques being used by both attackers and defenders. Besiegers used siege ramps, mobile siege towers, scaling ladders, rams and possibly mines whereas defenders are depicted firing or throwing missiles from city walls and employing an anti-ram device, catching it with chains and upending it. Both attackers and defenders used archery and incendiary devices such as pots of boiling oil and flaming missiles, making use of locally available sources of bitumen and oil. Two basic approaches to the capture of cities are illustrated: direct violent assaults on the fortifications and blockade, setting up a fortified camp outside the city walls and starving the besieged into surrender. These remained the basic methods and techniques of siege warfare, developed and evolved in the following centuries, and employed with varying degrees of success by other societies, but with particular success by the Romans **(49)**.

The Persians were the successors of the Assyrians in siege techniques and their efforts have been well illustrated by archaeological excavation at Palaeopaphos. The Greek island of Paphos had participated in the Ionian revolt against the Persians, who besieged the town in c.497 BC. They built a siege ramp, destroying a sanctuary outside the city to provide materials. Missiles found around the ramp probably represent attempts to halt its construction, and the defenders undermined the ramp, causing part of it to collapse. The Persians may have captured the city following a violent assault, but the archaeological evidence cannot provide these kinds of details. Siege warfare in classical Greece tended to be on a relatively small scale, but Plataea provides a good example of a blockade. In 427

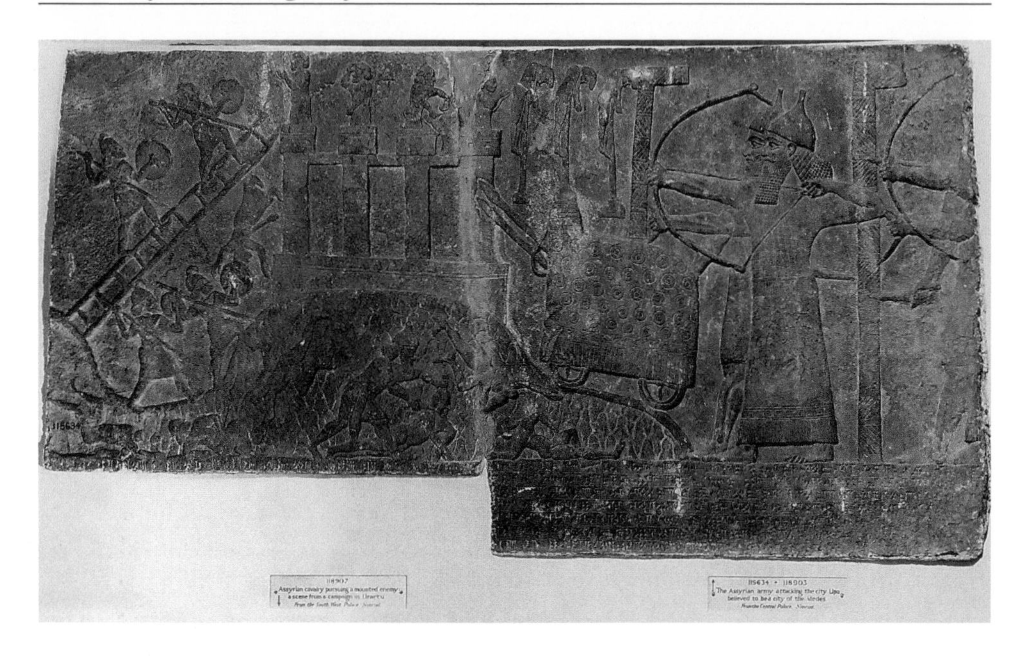

49 *Assyrian siege warfare: the techniques include scaling ladders, a siege engine, and covering fire from archers.* Courtesy The British Museum

BC, this town in Boeotia was besieged by Peloponnesian forces which tried a number of different methods to capture it, including building a ramp and direct assault, but the Plataeans heightened their walls and undermined the ramp. With the failure of direct assault, the Peloponnesians built a line of circumvallation with a ditch and two brick walls connected by a roof to provide shelter for the garrison; towers were set at regular intervals. The defenders were eventually forced by starvation to surrender.[144]

During the fourth century BC and the Hellenistic period the scale of siege warfare increased considerably, with the use of massive artillery and huge mobile siege towers, and with corresponding developments in the size and complexity of city fortifications. Torsion artillery, stone and bolt throwing machines, was allegedly invented for Dionysius I, the tyrant of Syracuse in the early fourth century BC, and came to be widely used in sieges for both offence and defence, particularly as an anti-personnel device. Its use in providing covering fire for Alexander's troops at Halicarnassus seems to have been a major factor in the success of the assault. The effectiveness of artillery depended on the skill of the architects who designed them and their operators, and textbooks were compiled containing full details and measurements for the construction of different pieces of artillery. Working examples of Hellenistic and Roman artillery pieces have been constructed by experimental archaeologists working almost exclusively from the texts, and their power and accuracy are very impressive.[145] Perhaps the most skilled military engineer of the Hellenistic period was Archimedes, working in Syracuse in the late third century BC. Before the Roman siege, the fortifications of Syracuse had been strengthened and redesigned to make the artillery a more effective defensive weapon. Archimedes had set up catapults of varying size and range to keep the Roman ships and soldiers away from the walls. Once the Romans were too close to the walls to be hit by the large stone and bolt-throwing machines, smaller engines at lower levels were able to target the enemy. Other machines caught and smashed the ships attacking the harbour area of Syracuse, frustrating any hope of a water-borne assault on the city. So exceptional were Archimedes' engineering skills that some accounts get carried away and even claim he concentrated the rays of the sun as a kind of ancient laser, but this is rather far-fetched. At any rate, his machines caused the Romans sufficient difficulties and casualties that they resorted to blockade and took the town in a surprise attack. Syracuse became one of the most famous sieges in the ancient world, not because of the strategic importance of its capture, but because of the image of one man, Archimedes, thwarting almost single-handedly the entire Roman army.[146]

Huge mobile siege towers were also fashionable in the Hellenistic period, most famously those built for Demetrius Poliorketes (the Besieger) at Salamis in Cyprus and at Rhodes. The historian Diodorus Siculus describes the Rhodian tower in detail but unfortunately, Demetrius had little success with his towers. The one at Salamis was burnt and he was unable to bring that at Rhodes into action properly, so their effectiveness cannot be judged; but they may have been too big and cumbersome for practicality. The construction of machinery like this, and indeed the waging of siege warfare as a whole, was a very expensive business. For the defenders, elaborate fortifications and preparations were necessary; for the attackers, supplies of food, water and the raw materials for the construction of siege works as well as the necessity to keep an army in the field, and static, for an unknown length of time. Not all societies or city-states could afford this kind of investment, or had the military organization necessary for success in this type of warfare. Most 'barbarian' armies, for example, generally did not possess sufficient organization or logistical support, tended not to remain in existence for very long and usually lacked the ability to bring a siege to a successful conclusion. After defeating the Romans in pitched battle at Adrianople in AD 378, for example, the Goths attempted to besiege the fortified city of Adrianople, but lacked the skills and tenacity necessary and wandered off in search of easier booty. The Romans, on the other hand, even in the Republican period when armies were not retained on a permanent footing, had the organization and logistical support necessary to prosecute a long blockade and attempt assaults with elaborate engines. They also, like the Assyrians, had the tenacity and determination to allow nothing to stop their progress. The militaristic nature and organization of Rome allowed her to be extremely successful at this type of warfare.[147]

Textbooks on siege warfare

Siege warfare is the subject of the earliest surviving military treatise from the ancient world, 'On the defence of fortified positions' written by Aeneas 'Tacticus' in the fourth century BC. Aeneas compiled a whole encyclopaedia of military matters, of which only the section on defending cities survives. The encyclopaedia included books on castrametation, planning, military finance, plots, and a companion volume to his work on defending cities that dealt with besieging them. Treatises written by others followed, including more specialized works on military machinery and technology. One of these, another encyclopaedic work by Philon of Byzantium dating to the third century BC, also covered various machines and devices, but contained a more general section, entitled *poliorketika* (*On siege warfare*) on attacking and defending fortified sites. These Greek works are comprehensive in that they deal with both attack and defence. The treatises dating to the early Roman Empire, however, tend to concentrate more on the offensive side of siege warfare. Although Frontinus includes stratagems on both attacking and defending fortified positions, Onasander does not consider their defence at all, and indeed this emphasis on attack is characteristic of his work as a whole. Mechanics and engineering in the ancient world came within the remit of architects, as did the construction of city walls, and so in his treatise on architecture, Vitruvius touches on certain aspects of siege warfare, particularly assault machinery, and when describing the construction of a town's fortifications he takes into account the possibility of a siege. His walls, towers and gates are designed with this in mind. However, he feels it unnecessary to say much at all about the defences of cities because, he claims, 'the enemy do not make use of our manuals when they besiege a city.' Vitruvius, it seems, expects only 'barbarians' to attack Roman towns, but his comment implies the practical value of Roman textbooks on this topic.[148] Apollodorus' treatise too is aimed at those besieging cities rather than defending them and the emphasis in these works reflects the more aggressive nature of Roman warfare in this period. In the later Empire, however, this trend is reversed and although Vegetius treats siege warfare in the same way as Philon, from the point of view of both attacker and defender, his emphasis is definitely on defence.

The technology of siege warfare

The expense of siege warfare encouraged attacking armies to attempt violent assaults; blockade usually occurred whilst preparations for assaults were made or if assault failed. The equipment available for such assaults is described in sometimes great detail in military textbooks, and by historians fascinated by the size and intricacies of the engines. Caesar and Josephus both describe Roman siegeworks in considerable detail, and Ammianus even includes in his history an excursus on the various types of siege machinery available in the fourth century AD. However, whilst the textbooks tend to concentrate on detailing the equipment rather than explaining how to execute a siege from start to finish, the historians illustrate in their accounts how the various devices for attack and defence could be used and countered. Scenes from sieges are also illustrated on public sculpture of the imperial period, such as Trajan's Column and the Arch of Severus in Rome, but these provide a snapshot of one point of the siege rather than a narrative.

Shelters

Shelters provided protection for the attackers approaching the walls of towns for assaults or for mining operations. They were generally made from a framework of stout timbers with planks and wicker hurdles on the sides and roof as protection against missiles. The whole structure would have been covered with uncured hides or some other fireproof material to protect it from incendiary devices. Fireproof coverings were particularly important because of the often extensive use of incendiary devices by both sides in siege warfare, but fire was one of the principal means of defence against all pieces of siege machinery. The different types of shelter are described by Apollodorus and Vegetius. Vegetius suggests different uses for the different types of shelter: the *testudo* (tortoise) shelter for example held the ram and there were specific shelters for sappers and archers, but the historical accounts of sieges do not make such distinctions. *Testudo* is the name given to both the shelter holding the ram and the formation made by Roman soldiers from their large shields. Both are tortoise-like, and Vegetius explains that the ram shelter gets its name because the head of the battering ram swings in and out of the shelter like the head of a tortoise from its shell. Artillery and incendiary devices were the recommended defences against these shelters, and at Marseilles during the Civil War, Caesar's soldiers were forced to work behind shelters made of four layers of hurdles because of the power of the Masilliotes' artillery, and even these were pierced. Such was the strength of the artillery in Marseilles that Caesar's troops were forced to build a gallery from brick and timber to allow them to get up to the walls, and a brick tower to lay down covering fire.[149]

The *testudo* formation made by soldiers from their shields is described by both Livy and Cassius Dio. Livy suggests that the formation was also used in a performance act in which young men formed a tortoise with their shields and two more fought a mock battle on top of the structure:

> (About sixty armed men) would form closely packed ranks, with their shields packed together above their heads. The first rank stood, the second crouching a little lower, the third and fourth even lower, and the final rank kneeling, and together they formed a tortoise with a slope like the roof of a building...
> A tortoise like this was moved up to the lowest part of the wall (at Heracleum). When the soldiers approached the wall on the tortoise, they were level with the top of the wall; the defenders were forced down from the walls and soldiers from two maniples climbed over into the city. The formation was the same as in the show, except that those in the front rank and on the flanks held their shields not above their heads, but in the normal position for fighting, so that they did not leave themselves open. Thus, the missiles hurled from the walls did not harm them as they approached, and those thrown onto the tortoise slid like rain harmlessly down the slippery slope.
>
> Livy 44.8

In Livy's military show, two soldiers performed on top of the *testudo* but Dio claims the formation could support horses and vehicles. Dio's formation does not have the incline of Livy's but this refinement was only necessary in siege warfare to keep the missiles off or to allow attackers to climb up it. The other role of the *testudo* formation was in open warfare when a Roman force of heavy infantry was faced with a mobile army strong in archers, and in this case probably all the soldiers would have remained upright. The formation is recommended for this purpose by Onasander and Dio's description comes at the moment that Antony deploys it to counter the Parthian archers. The *testudo* is illustrated on both Trajan's Column and the Column of Marcus Aurelius. In the latter example, various missiles, including rocks, firebrands, and what appears to be a pot of

50 Testudo *formation, Column of Marcus Aurelius.* Courtesy DAI

boiling liquid, roll harmlessly off the shields, just as Livy claims in his description **(50)**. The *testudo* allowed troops to get up to a wall, to undermine it, to assault it by climbing up the shields, or to make a forced entry into a town under its protection. The first hillfort or *oppidum* that Caesar captured in Britain was taken by soldiers piling up earth against the fortifications under a *testudo* formation until it was high enough for them to climb over the defences. Vespasian's troops attempted to enter Jotapata over a boarding bridge under cover of one, but Josephus tells with evident delight how he broke up the formation by pouring boiling oil on the soldiers. When they had re-formed for a second attack, he poured boiled fenugreek onto their gangways so they all lost their footing.[150]

Siege towers

The construction of siege towers is covered by Apollodorus in some detail, and by Vegetius more briefly, whilst Ammianus includes a description of the tower in his excursus on siege equipment. The details of siege towers provided by historians in their accounts of sieges indicate their interest in these large and very elaborate machines. A siege tower could contain a number of different devices, including artillery at all levels, rams, swing-bridges and boarding ramps to allow the attackers access to the walls. Its very size and height would have given cause for concern and just the construction of one might

51 *Siege tower with battering ram in action in Parthia, early third century AD; Arch of Severus, Rome.* Courtesy The American Academy in Rome

have encouraged the defenders to consider surrender, as the Suessiones did at Noviodunum.

Josephus reports that the Roman siege towers at Jotapata were 50 feet high and encased in iron plates to protect them from fire, whilst that at Masada was 75 feet high. Aquileia was protected by its position on a river, so the forces of the Caesar Julian in AD 361 built siege towers on wooden platforms formed of ships lashed together. Soldiers on top of the tower kept the defenders off the wall whilst light armed troops at a lower level crossed on gangways and tried to breach the wall. However, the attackers seem to have failed to take the necessary precautions against fire, and the towers fell victim to incendiary devices. Normally siege towers would be protected from fire by the iron casings Josephus mentions at Jotapata and Masada, or by the same methods advised for shelters. Apollodorus suggests an additional device, a kind of sprinkler system fed by tubing made out of ox's intestines and siphons which carried water to the top of the tower, but there are no reports of such a system actually being used.[151]

The principal sculptural reliefs on the Arch of Severus in Rome illustrate sieges from the emperor's campaigns in Parthia in the late second century AD. In one scene a tower is shown with a ram on one of the lower levels and troops on the top levels waiting to storm the city, whilst others dig a mine into the city **(51)**. The siege tower at Masada, by contrast, had its battering ram on an upper level as well as, presumably, its boarding bridge. The technical treatises do not specify at what level of the tower items of equipment should be placed: towers were constructed for a specific siege and although the overall design might not vary much, the details were devised for the requirements of that particular siege and could therefore vary considerably.

Vegetius advises several defences against the advance of siege engines; sections of the wall could be strengthened or heightened to prevent siege towers dominating the walls, though he notes that besiegers sometimes built a tower with another turret inside it that could be suddenly raised by ropes and pulleys to over-top the wall. Mines too, he suggests, could be dug where the tower would approach so that it would subside before it reached the walls. When a tower was moved up, the defenders could attempt to drive it away from the walls using long wooden beams bound with iron for extra strength. Onasander, concentrating on the art of attacking a city rather than defending one warns the general about the dangers of sorties, and they occurred often in sieges, meeting with varying degrees of success. At Athens and Marseilles, the defenders were able to burn the siege engines, though in the latter case it was during a truce so Caesar's soldiers were slow to react. At the siege of Amida in AD 359, however, the defenders got into difficulties during over-optimistic sorties and eventually the gates were blocked up, to prevent sorties rather than to provide additional security.[152]

The strengthening and heightening of walls in preparation for an attack or during a siege was a common practice. At Dura Europos on the Euphrates, long earthen mounds were built by the defenders inside the walls to strengthen them against attack from rams, and outside for additional protection and to hinder the approach of siege towers. Walls were strengthened at Veii in the late fifth or early fourth century BC, probably because of the prospect of a Roman siege, at Syracuse by Archimedes, as mentioned above, and at Cremna, a Pisidian rebel's fortified town. On other occasions, walls were heightened in reaction to the attackers' operations, at Avaricum, Jotapata, Dura Europos and Amida. The undermining of siege ramps seems to have been common practice from comparatively early, as the archaeological evidence from Palaeopaphos indicates. The Gallic defenders tried to undermine and fire Caesar's huge siege ramp at Avaricum, without success, but efforts by the Jews at Jerusalem and Romans at Dura Europos did succeed in preventing the approach of siege engines. Long poles of the type suggested by Vegetius were used by the Roman defenders at Vetera in AD 69 when the legionary fortress was besieged by the revolting Batavians. The two-storey tower they had built was demolished by blows from the poles, but the construction may not have been very solid: it had been built by Roman prisoners because the Batavians were unskilled at siege warfare.[153]

Battering rams

The fundamental principal of the battering ram is very simple, but the effectiveness of the ram could easily be greatly increased **(52)**. The ram is described in detail by all the treatise writers that touch on siege warfare, no doubt because of its usefulness and ubiquity. The ram would usually be suspended in a mobile shelter for protection. Apollodorus states there should be chocks by the wheels to fix the shelter to the ground when the ram was in use, so that the axles did not take the weight of the moving equipment, and to prevent

52 *Battering ram in an assault on a Persian city, early third century AD; Arch of Severus, Rome.* Courtesy The American Academy in Rome

it skidding. That would provide greater strength to the blows of the ram. A ram in several pieces should have several hanging points to retain its strength and solidity. He also recommends use of the ram against gates, angles of towers and any other weak points in the walls.[154]

Rams could vary in strength, and two noted for their particular power are the Roman ram at Jerusalem that the Jews nicknamed Victor and a ram used by the Persians against the Roman held town of Bezabde in AD 360. Ammianus claims it was a hundred years old, and it was made from several pieces, presumably for ease of transportation, but this does not seem to have affected its strength. Whilst it seems likely that reconnaissance would have been made to discover weak sectors of the walls against which the ram would have had more effect, only Ammianus specifies this. At Singara in AD 360, the Persians used the ram against still damp mortar in the joints of newly built walls and forced an entry at this point, whilst at Bezabde they attacked parts of the walls that were unstable and falling down.[155]

Vegetius makes several suggestions for countering rams. Heavy missiles, even pieces of column, should be thrown or rolled down the walls onto the shelters and rams in an attempt to break off the heads; padded mats could be lowered to cushion the blows of the ram; or the defenders could catch the ram in a noose and haul it up the wall. These are obvious suggestions to make, and such defences are illustrated on the Assyrian reliefs mentioned above. Missiles thrown down from the walls often included chunks of masonry, millstones or column drums; the defenders at Syracuse rolled missiles down the cliffs as well as using artillery, and a very large rounded stone from Cremna may have been used for this purpose rather than as artillery shot. At Haliartus in Boeotia in 171 BC, the defenders used stones and lead weights to destroy the Roman rams whilst in 189 BC at Ambracia, cranes were used to drop similar weights. Even if the missiles missed the rams, they would still be a dangerous diversion to those working the ram. Josephus was able to protect the walls of Jotapata for a while using sacks filled with chaff, which Vegetius also suggests, until the Romans cut them down using reaping hooks attached to long poles. The Massiliotes, who caused Caesar's army considerable difficulties during their assaults on Marseilles, caught one of the rams in a noose and wound it up with a windlass.[156]

The battering ram, though, was not just a device for creating breaches in walls; it seems to have had a greater significance in Roman warfare. The tribe of the Atuatuci were besieged in their hillfort (or *oppidum*) by Caesar during his campaign of 57 BC, and made enquiries about terms of surrender. Caesar's response was that he would be merciful provided the Atuatuci surrendered before the battering ram touched the walls of the *oppidum*. Cicero advocates that mercy should be shown to those who surrendered 'even though the battering ram has hammered at their walls'. The moment at which the battering ram came into action in a siege may have indicated the point at which the siege proper began, and this had implications for the treatment of the defenders, as Caesar and Cicero's comments suggest. Josephus mentions the despair of the Jews when the Romans brought their rams to bear on the walls of Jerusalem. This was probably because it was the moment that the siege began in earnest, and there was no turning back.[157]

Mines

Much advice is given in the treatises about the construction of mines by both attackers and defenders, and on how they could be countered. Mines could be used to enter a city secretly and capture it, as the Romans did at Maozamalcha in AD 363, and are supposed to have done at Veii in 396 BC. Tunnels were more usually dug to undermine city walls. The sappers would then underpin the walls with wood and fire the supports with flammable materials such as resin and sulphur, causing the walls to collapse. Vegetius suggests that cities should have a wide deep ditch outside the walls to prevent any mining attempts as well as to hinder the approach of siege engines, but he provides no other advice on countering mines, and neither does the offensively-minded Apollodorus. Vitruvius, who was an artillery officer under Caesar and was probably in his army at Marseilles, describes how the defenders there had created a water-filled basin inside their walls when they noticed mines were being dug. When they reached the basin the mines were flooded[158].

The Roman mine workings at Ambracia in Greece had been noticed because of the spoil heaps, despite attempts to hide them. The defenders then dug a trench inside the wall and hung on the side bronze plates to detect the location of the tunnel through the vibrations, a method of detecting mining operations which is explained by the Greek military writer Aeneas Tacticus. A counter-mine was dug to intercept the Roman one but by this time, the Romans had underpinned a large part of the wall. When the tunnels met an underground fight broke out, firstly between the sappers using their tools as weapons and then between soldiers sent in by both sides. The battle seems to have gone on for some time but then slackened as sappers blocked off the tunnels with cloths and wooden trap doors. The Greeks then flushed the Romans out of the tunnel using a device to pump noxious fumes through the mines, forcing the Romans to retreat. The Romans besieged at Dura Europos similarly dug a counter-mine to try to prevent the Persians destroying a section of the wall that they had already undermined and underpinned. As at Ambracia, an underground skirmish occurred when the two mines met, with the Persians coming off better. The Romans seem to have panicked at some point and, fearful that the Persians would actually enter the city through the tunnels, blocked their counter-mine with rubble. Some 16 to 18 Roman soldiers who were unable to get out of the mine in time were trapped there, between the rubble and the Persians. The latter then withdrew and fired the mine, causing it to cave in, and killing the trapped Romans. Although the Persians fired the mine with pitch, straw and probably sulphur, the wall slumped into the mine and was not destroyed, leaving the Persians to find another way into the city **(53)**.[159]

Wall breaches

> But if the force is so great that as often happens the wall is pierced by the rams, one hope of salvation remains. That is to demolish houses and build another wall inside, so that if the enemy tries to get through he will be cut off between the two walls.
>
> Vegetius 4.23

What Vegetius fails to point out is that the soldiers attacking the inner wall in this situation would be exposed to artillery fire or other missiles from the flanks as well as the front. This happened to Sulla's soldiers at Athens, and the fire was so heavy that he was forced to retreat. When he later renewed his attack on this section of the wall and succeeded in breaking it, he found that the defenders had anticipated this and built several others like it. This would most likely have had a detrimental effect on the morale of those attacking as well as costing time, resources and men. When the Romans encountered this defence at the fortress of Antonia in Jerusalem, Titus' soldiers were extremely reluctant to attempt to storm it. Titus was forced to make a personal appeal to his men, promising promotion to the man who captured the wall. Only a handful of the soldiers attempted an assault and when all were killed or wounded the others made no further concerted effort to capture the wall because of the intense danger. The fortress was eventually taken by a 'stratagem' when a small group of Romans entered the fortress secretly and sounded their trumpets; the main Roman force attacked at this point but the Jews fled, believing the fortress already captured.[160]

It is probably because of the dangers involved in storming walls that a decoration in the form of a gold walled or mural crown was awarded to the first man who captured the wall of an enemy city. It is clear that there could be intense competition for the award of this crown, the *corona muralis*, as Livy reveals in his account of the capture of New Carthage in 210 BC. Two men, a legionary centurion and a marine, claimed the prize for being the first over the wall and a near-mutiny arose from disputes between the legionaries and marines supporting their own candidate. Eventually both men were awarded a crown. At New

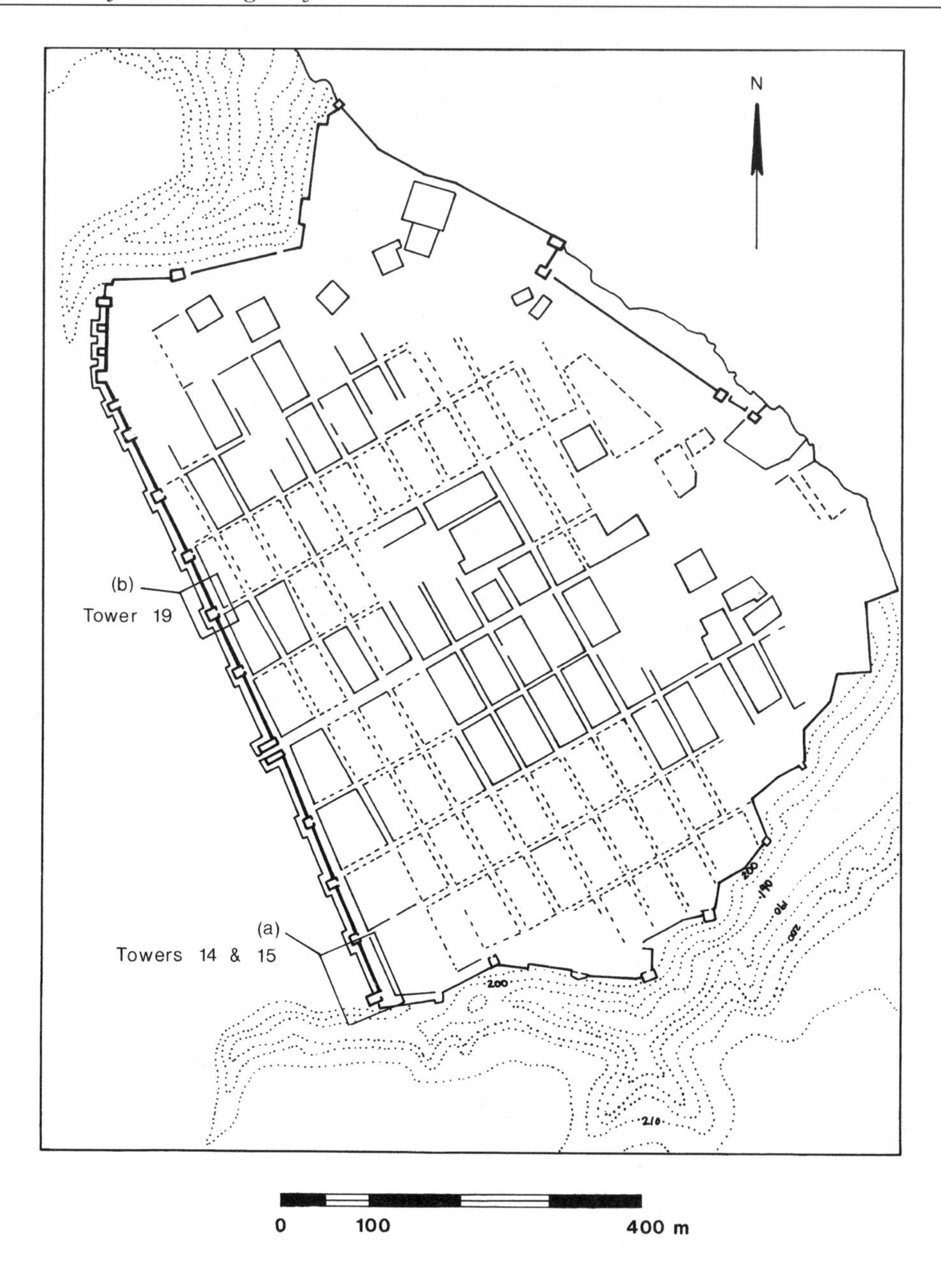

53 *Third century AD Parthian siegeworks and Roman counter measures at Dura Eurpos on the Tigris. The Parthians assaulted the more vulnerable western wall of the city, but probably came under fire from artillery on the towers. At the south west corner of the city, the attackers built a siege ramp, and undermined Tower 14 (a), possibly to prevent its use as an artillery platform firing on those building the siege ramp. The Romans attempted to undermine the ramp, but their mine wandered off course. The undermining of Tower 19 and the wall by the Parthians had been disrupted by the construction of the Roman counter-mine, but failed to create a breach as the wall slumped into the mine and remained upright (b).*

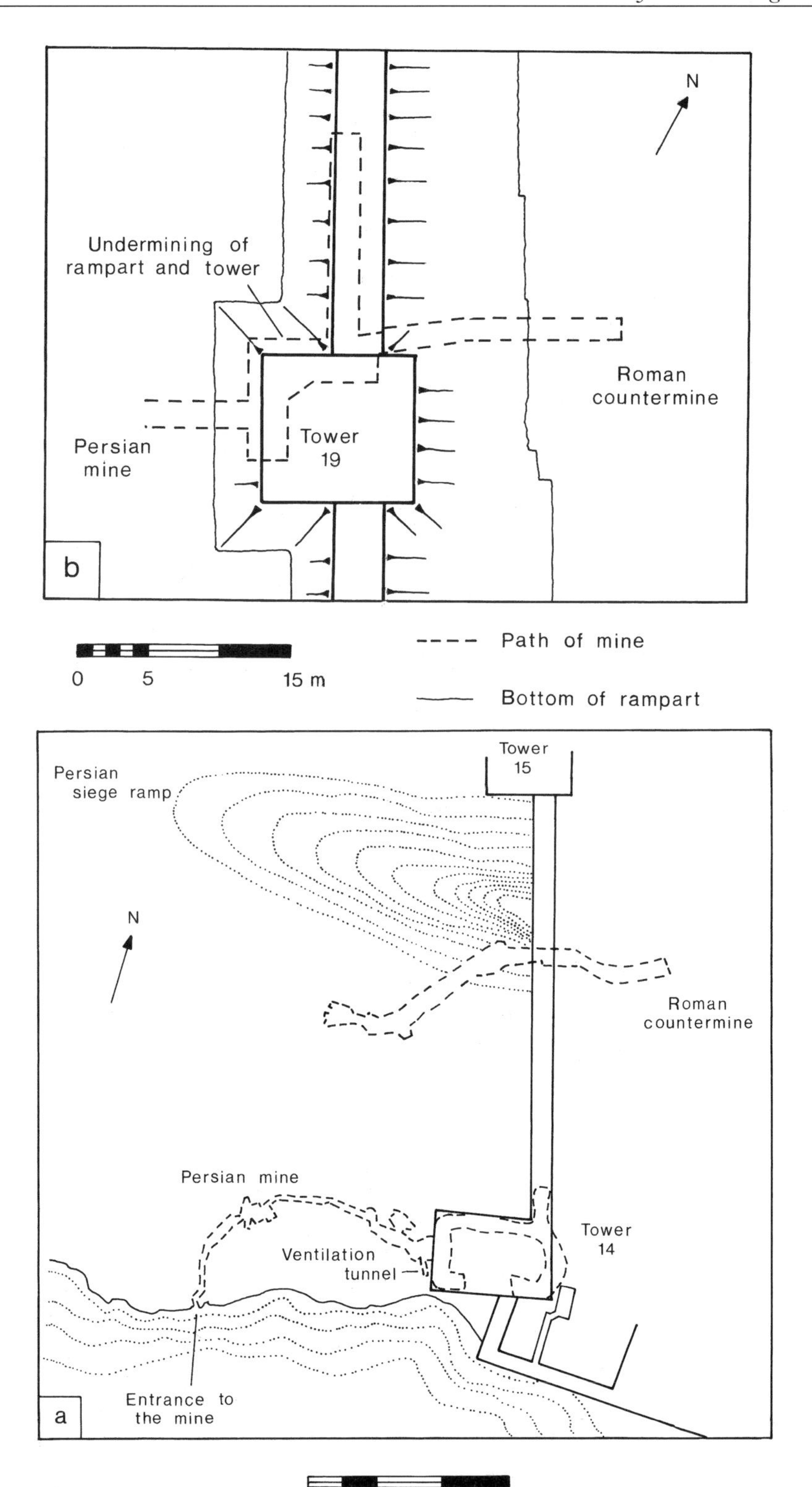
N
Undermining of rampart and tower
Roman countermine
Tower 19
Persian mine
b
0 5 15 m
Path of mine
Bottom of rampart
Tower 15
Persian siege ramp
N
Roman countermine
Persian mine
Ventilation tunnel
Tower 14
Entrance to the mine
a
0 5 10 30 m

Carthage the Roman soldiers were eager to compete for the decoration and to face the dangers involved in winning it and this encouraged them in the assault. Sallust too claims that men once competed with each other to be first to storm the rampart, only for the honour such courage brought, compared with the moral decline he perceived in the late Republic. In the imperial period, the *corona muralis* was only awarded to those of or above the rank of centurion and so at Jerusalem Titus had to offer other incentives in the form of promotion to encourage his men to storm the walls. It seems clear, however, that Titus was asking too much of his men in this situation and their loyalty to him and desire for promotion did not, for the most part, encourage them to an attack that was foolhardy, even suicidal.[161]

Artillery, missiles and incendiary devices

> I shall leave out military machines and engines, the invention of which has already reached its limits, and I see no further development in this area.
>
> Frontinus, *Stratagems* 3 preface

Frontinus proved wrong in his prediction, with developments in both bolt and stone throwing engines, but these matters are dealt with in detail elsewhere, and we shall concentrate here on the use of artillery and incendiary devices in siege warfare.[162] Although various treatises explaining the construction of artillery pieces do survive from antiquity, they tend not to provide practical advice on using the weapons under particular circumstances. It is mainly from Vegetius and descriptions of sieges in histories that we can obtain this information.

Vegetius states that the principal use of artillery in attack was to keep the defenders off the walls, and he later notes the defensive role of artillery against both men and machines. Since its introduction in the fourth century BC, artillery was used as an anti-personnel weapon: Marcellus had placed artillery on his ships at Syracuse to provide covering fire for his troops, though the ships had been demolished by Archimedes' powerful stone-throwers. Roman artillery provided covering fire for the troops building a siege ramp at Jotapata and Josephus claims that one particularly effective stone-throwing *ballista* knocked off a man's head and flung it more than 600 yards, a far less likely claim than his observation that the machines could knock off battlements and the corners of towers. Ammianus provides particularly dramatic information about the effectiveness of artillery at Amida. The Persians placed pieces in their siege towers to drive the defenders off the walls, but Roman onagers were able to destroy the towers. So powerful were the Roman bolt throwers that when some troops got into trouble on a night sally and fled back into the city, the Romans fired the artillery blank, without missiles. Just the sound of the catapults firing caused the Persians to hold back their pursuit, allowing the Romans to reach safety. Josephus too talks of the sound of the machines when they fired, the rushing sound of the missiles and the final crash when they hit something. Missiles from the stone-throwing ballistas and onagers might be seen. At Jerusalem, the Jews were able to spot the white stone missiles and warn their colleagues to take cover. The Romans then painted their missiles black so they were harder to spot. The bolt throwing machines

54 *A bolt-shooting machine, known as the* scorpion *because of its deadly sting.* Courtesy JCN Coulston

might be heard, but the missiles would be almost impossible to see and in any case would be travelling too fast for evasive action. The psychological effect of these weapons was not insignificant.[163]

In addition to artillery, slingshot of lead or clay was an effective anti-personnel and incendiary device. Bullets have been found at a number of sites dating from the first century BC to the third century AD, but they were in use well before this period. Two sites in Italy are of particular interest: Asculum, besieged by Pompeius Strabo during the Social War between Rome and her Italian allies, and Perugia, blockaded and assaulted by Octavian in 41 BC when it was held in an insurrection led by Mark Antony's brother Lucius. Many of the lead shot at both these sites were inscribed with the names of legions, officers, commanders, and derogatory comments and illustrations as part of an airborne propaganda campaign. Such graffiti could be more useful though; Appian reports that when Sulla was besieging Athens and the Piraeus in 87 BC, two Athenian slaves started sending information to him by inscribing messages on lead slingshots and firing them at the Romans. Because of information received in this way, Sulla was able to ambush and capture a supply train.[164]

Clay slingshot, when heated, was an effective incendiary missile. It was used by the Gallic tribe of the Nervii to burn the buildings in their attack on Quintus Cicero's winter camp in 54 BC, but there were other more elaborate incendiary devices available. Artillery and archers as well as slingers could shoot incendiary missiles, or they could simply be

55 *Roman siegeworks at Burnswark, S.Scotland. The* tituli *of the Roman camp in the foreground doubled up as artillery platforms.*
Courtesy Cambridge Committee for Aerial Photography

dropped on attackers. Vegetius advises cities to prepare tar, sulphur, pitch and oil for burning the enemy's siege engines. We have seen how Josephus broke up a Roman *testudo* formation with boiling oil, whilst at Uxellodunum the Romans had to cope with burning barrels rolled down on them by the Gauls. In the Eastern parts of the Empire, incendiary missiles were particularly common and unpleasant, because of the easy availability of pitch, bitumen and naphtha. Burning naphtha was almost impossible to extinguish and easily burned up anything it touched, man or engine. It is perhaps no wonder that Severus' soldiers at Hatra mutinied and forced the emperor to abandon his siege of the city with its strong defences, artillery and incendiary missiles.[165]

Vegetius mentions two types of incendiary missile, the *malleolus*, shot by an archer, and the *falarica*, fired by artillery. Ammianus describes the *malleolus* and advises that it should be shot slowly from a loose bow to keep the fire alight. He does not mention the artillery version, the *falarica*, but Vegetius says it could pierce solid iron or the protective layers of a siege engine, allowing the burning material to reach the wooden structure.[166] Historians tend not to be so specific in their terminology, but do mention both types of incendiary missiles in various sieges ranging in date from Ambracia in 189 BC to the wars between Rome and Persia in the mid fourth century AD. According to Appian, the Romans besieging the town of Delminium in Dalmatia fired into the town wooden shafts two cubits long, covered with flax, pitch and sulphur, using catapults. As with other types of missile, the incendiary missiles were used against both personnel and machines, and could understandably be particularly effective in naval warfare. Both sides at Actium in 31 BC fired pots of charcoal and pitch at each other's ships by means of artillery. Perhaps more spectacular is the naval battle Hannibal fought for Prusius of Bithynia against the Rhodians in which he fired from artillery terracotta pots filled with poisonous snakes. The pots broke on impact, sending snakes all over the ships, and the enemy fled. Herodian claims that the defenders of Hatra filled clay containers with flying insects with poisonous stings and fired these at the Roman besieging force, compelling them to retreat. The story is not entirely trustworthy, but is typical of the attractiveness of such cunning stratagems to historians writing accounts of sieges.[167]

The siege

No one ancient textbook explains how to carry out a siege from start to finish, but the information in textbooks and historical narratives allows us to study the different methods the Romans used in siege warfare. There were two ways of attacking a fortified position, by direct assault or blockade. The first could be sufficient to capture the objective, but if the risk involved in such a venture were too great, a stronghold might instead be blockaded to induce surrender through starvation. Scipio decided to do this at Numantia, because his army was demoralized by previous defeats. Although he had put his soldiers through a rigorous training programme, he wished to avoid direct conflict with the Spanish who, according to Appian, would fight with the strength of desperation. This kind of strategy was rare because of the length of time such a siege would last, time during which the besieging army was also likely to experience problems in the supply of food and water when it had consumed any local foraging. A more common approach was to blockade a stronghold to weaken and demoralize the enemy whilst preparations were made for an assault with siege machinery.

Assaults

Onasander recognized the importance of speed in siege warfare. His advice is more concerned with the violent assaults of cities than with blockades, and he suggests that the general and his officers should encourage greater effort from their soldiers by getting involved in the work themselves. The presence of the commander during such work could enthuse the soldiers, as Titus did at Jerusalem. Competition between legions and cohorts, and their officers, ensured that the construction of the circumvallation at Jerusalem was completed quickly. Titus' desire for speed though was encouraged not by the cost of siege warfare, according to Josephus, but by a concern that the glory of his triumph in capturing Jerusalem would be diminished if the campaign were drawn out. Reputations, says Josephus, were won by speed. Onasander suggests that the general should make a sudden assault to give the defenders less time to react, dividing his army to weaken the besieged by continuous attacks at different parts of the walls. Feint attacks too could divide and confuse the defenders. Vegetius warns that this initial assault is frequently the most dangerous for defenders because of its violence and the determination of the attackers. Advice on countering such assaults is very limited. Vegetius suggested building up supplies of missiles in preparation for a siege, and these could be used against the initial assault.[168]

> After inspecting the defences and making preparations for the assault, Corbulo encouraged his men to throw out this shifty enemy who wanted neither peace nor war, but had revealed their treachery and cowardice by fleeing, and to seek glory as much as booty. Then he divided his army into four sections. One he ordered to mass together in a tortoise formation and undermine the wall. Another he ordered to move ladders up to the walls, a third large section to shoot firebrands and javelins from the artillery. He assigned a position to the slingers and stone shooters from which they could fire missiles from a distance, so that the enemy would be hard pressed at all points and would be unable to help those in trouble. The assault was so keen that before the day was one third over the walls had been stripped of their defenders, the gates overturned, the fortifications taken by storm, and all the adult males slaughtered; all this without a single Roman soldier lost, and with only a few wounded.
>
> Tacitus, *Annals* 13.38

Corbulo's attack on the Armenian city of Volandum in AD 58 provides an almost textbook example of the type of violent assault recommended by Onasander. Having made a reconnaissance, Corbulo divided his force; whilst half provided a highly effective covering fire from artillery and slings, the other half attempted to undermine the walls under the protection of a *testudo* formation, and to scale them with ladders. The defenders were overcome by the violence of the attack, and Corbulo took the city in a few hours with no casualties. Caesar's very swift attack on the city of Gomphi in Thessaly had similar results. Such attacks not only saved time, resources and lives during a campaign, but their speed and ferocity might give the general a psychological advantage; other towns might be shocked into surrendering, though there were other factors which might help to encourage such action. After the brutal treatment of Gomphi, the remaining towns in Thessaly that had originally sided with Pompey opened their gates to Caesar.[169]

Blockade and circumvallation

Traces of a circumvallation are often one of the few physical features of a siege to survive. Caesar's double line of fortifications at Alesia is well known both from his own description of the works and from the excavations conducted for Napoleon III in the nineteenth century. Traces of much less elaborate stone circumvallation walls have been identified at Machaerus and Beththera in Israel. Machaerus was besieged by the Romans in the Jewish revolt of AD 66-73, Beththera in that of AD 135, and although there are literary accounts of the sieges, none of them makes any reference to a physical line of circumvallation. This omission in the literary accounts is slightly surprising because the construction of a circumvallation suggests large-scale siege operations were being undertaken, but together the archaeological and literary evidence can supply valuable information about siege warfare.

Circumvallations are mentioned by Apollodorus and Vegetius, the former advising a rampart and ditch at least five feet deep. Vegetius sensibly notes that the line of circumvallation should be beyond weapons range of the besieged town, but he also seems to imply that this type of physical barrier was no longer used by besieging forces, and accounts of sieges after the early third century generally do not mention the construction of such fortifications. In the sieges of the mid fourth century described by Ammianus, blockade was imposed simply by the enemy encamping outside, but not pressed as closely as was possible with a physical barrier. This made the besieging force more vulnerable to sorties by the defenders, which is why Vegetius recommends the attackers build a circumvallation wall. This was not the only reason, however, for surrounding a town with a physical barrier. It would signal the intent of the army not to depart until the place was taken; it would emphasize the isolation of the besieged, cut off from supplies and a relieving army, if one was being raised; and it would make communication with the world beyond the siege lines extremely difficult if not impossible. Frontinus includes a number of stratagems illustrating the sending of messages to and from besieged cities, including how the consul Aulus Hirtius used carrier pigeons to get messages to Decimus Brutus when the latter was besieged in Mutina by Antony in 43 BC.[170]

The Roman circumvallation of Jerusalem, some 4½ miles (7km) long and including 13 forts, took only three days to complete, according to Josephus, and the speed of the work was likely to have increased the fear of the besieged. He also notes that the Roman siege works stripped the countryside of all its timber for a 10-mile (16km) radius round Jerusalem. Like the destruction of Jerusalem after its capture, this devastation to the countryside would have left a lasting scar on the Judaean landscape. Lines of circumvallation usually made good use of the local topography and were usually well out of reach of artillery range of the besieged town. This would suggest that Vegetius is right when he claims that the fortification of the investment line was primarily to protect the besiegers from sorties. At a number of sites, including Alesia and Masada, the distance between the circumvallation wall and besieged town was increased when the ground between them was open and more easy, giving the besiegers more time to react in case of a sortie **(56)**.[171] Once the ring of fortifications was complete the besieging army might turn to the construction of other siege works such as the towers, mines and ramps discussed above or, more rarely, wait until those within were compelled through starvation to surrender.

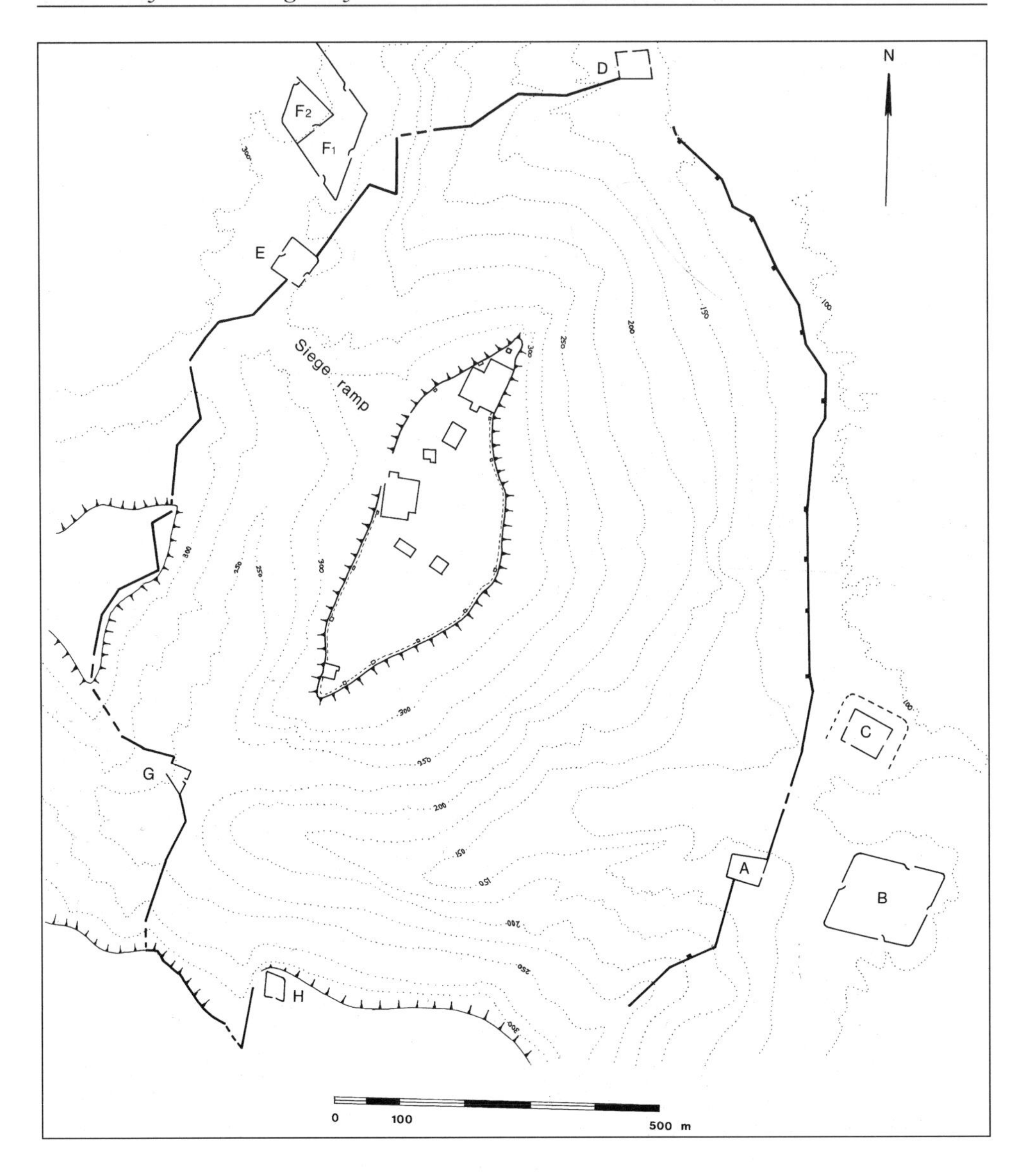

56 The Roman circumvallation and siege ramp at Masada. The circumvallation takes advantage of the terrain, and towers were only necessary on the less severe slopes of the eastern side

Supply and siege warfare

Vegetius considers in some detail the importance of preparing cities for withstanding a siege. The inhabitants should stockpile the materiel of war, including bitumen, sulphur, pitch and oil for making incendiary devices, missiles for artillery, rope and sinews for catapults (though women's hair was suitable in an emergency), and horn and hides for making engines and armour. They were also advised to ensure access to a good water supply, ensure supplies of salt, build up stocks of food, wine and fodder, and to grow crops in open spaces within the defences. Vitruvius too lists foods that can be stockpiled in case of siege, but stresses the importance of collecting firewood, and suggests that city colonnades should be used to store it. Vegetius also notes that non-combatants might be excluded from a city in order to preserve the food supplies for those capable of fighting.[172]

Decimus Brutus prepared Mutina against siege by Mark Antony in 43 BC by stockpiling supplies from the town's territory and slaughtering and salting livestock, no doubt one of the reasons that Vegetius advises ensuring supplies of salt in preparation for a siege. Josephus too seems to have prepared the towns of Galilee against siege by the Romans during the Jewish revolt. He had fortified and strengthened the defences of a number of towns, and had presumably organized collection of supplies, since he reports that Jotopata was plentifully supplied with corn and other necessities, apart from salt. Water supplies were a concern though, and Josephus had begun rationing it from the very start of the siege. Towns with a good internal water supply were obviously much more secure than those relying on rivers or aqueducts which could be diverted or cut. Frontinus devotes a section of his *Stratagems* to these sorts of schemes, including the poisoning of water supplies which was clearly considered a valid tactic. That it was normal to gather supplies in anticipation of a siege is indicated by Livy's explanation for the surrender of the Roman garrison of the Illyrian town of Uscana to Perseus in 169 BC. The Romans surrendered because it was becoming clear that Perseus' forces were likely to take the city by assault, but also because there were no supplies in the town, not even of grain, as the siege had been unexpected. Tacitus points out that the Roman general Paetus had surrendered to the Parthian king Vologeses in AD 62 despite being so well supplied that he had burned the granaries, and indeed the Parthians had been about to abandon the siege because they had run out of supplies and forage.[173]

Supplies could be a problem for those besieging, as the case of Vologeses shows. The Roman army at Jerusalem experienced water shortages, and although Josephus does not mention any difficulties in supplying the army besieging Masada, they must have been considerable. Whereas the small Jewish force in the fortress was well supplied with both food and water, the Romans had to obtain food from a considerable distance because the desert allowed no foraging; even water had to be transported from some miles distant. The major problem for a besieging force was that all local supplies and forage would have been exhausted very quickly, especially if the operation involved a large army such as that at Jerusalem, and supplies would have had to be brought in from further away. It was all the more important in such circumstances that the communications between an army's supply base or bases and its siege camps were secure and that supply trains arrived regularly to ensure the army could continue its operations. Such logistical concerns for a besieging army were likely to be more serious if the objective was inland, since the blockade of a coastal site could be more easily supplied by sea (though so could the besieged). The successful Roman sieges of sites like Jerusalem and Masada indicate the effectiveness of their supply operations.[174]

Because towns were likely to have gathered supplies in case of siege, they could be valuable to the besieger in providing him with food and other materials, as Scipio found on the capture of New Carthage in the Second Punic War. Metellus had clearly expected to re-supply his army from capturing Langobrigae, and Caesar may have decided to capture the town of Gomphi specifically to alleviate his army's supply problems. He had been forced through lack of supplies to abandon his blockade of Pompey at Dyrrachium and withdraw into Thessaly, but had been unable to re-supply his army, which was being weakened by hunger. The assault and rapid capture of the well-stocked town eased these difficulties, and had the more wide-ranging result of encouraging all the other towns in the region to defect from Pompey.[175]

Towns with a poor internal water supply, or none at all, were obviously extremely vulnerable to siege warfare since they would be unlikely to be capable of holding out for very long. Caesar deliberately attacked the water supply of the Gallic *oppidum* of Uxellodunum in 52 BC to force the defenders to surrender. Firstly he posted slingers, archers and artillery to prevent the Gauls from reaching the river at the foot of the *oppidum's* defences, reducing them to one spring only for their needs. An artillery tower covering the spring hindered access to the remaining source of water whilst Caesar's men dug a tunnel to divert the spring. Hirtius reports that the Gauls were already weakened by thirst when they made an all-out attack on the Roman siege operations, and were repulsed. They were eventually forced to surrender when the Roman tunnels diverted the streams that fed the spring, making further resistance impossible. This successful attack on the water supply saved Caesar and his soldiers from the expense and potential risks of an assault on the *oppidum*. In the campaigns against Sertorius in Spain, Metellus decided to capture the town of Langobrigae when he learned that it had only one source of water within its walls. The Roman general was convinced that the defenders would be unable to hold out for more than a couple of days. Metellus' intentions of besieging Spanish strongholds were more than once, though, thwarted by the intervention of Sertorius. Plutarch reports that whenever Metellus began a siege, rather than trying to raise it by force and risk battle with the Roman troops, who had the advantage in open engagements, Sertorius targeted Metellus' supply lines, forcing him to abandon his sieges. Sertorius' strategy shows well how vulnerable any army could be if its supply lines were not secure; attacks on the vulnerable supply train, or on foraging parties could prove disastrous, as Metellus' abortive siege of Langobrigae illustrates.[176]

Certain that the town would be unable to sustain a siege, Metellus ordered his men to take with them only five days' rations (presumably he was marching fast and light, detached from any organized and secure supply lines). If the Romans had taken the town as quickly as Metellus had anticipated, they would have been able to replenish their supplies from those captured. But Sertorius arranged for water skins to be smuggled into the town, and for the non-combatants to be guided out to preserve the supplies for those fighting (one can only assume that the Roman blockade was rather poor). Unable to capture the town as expected, Metellus was forced to send out a foraging party 6,000 strong, which Sertorius ambushed. Metellus wisely abandoned the siege.

Although supplies then might be of concern to a besieging army, and on occasion even force it to abandon the siege, more often of course, it was those enclosed within the circumvallation that suffered the shortages. Ammianus describes how the population of Amida had been greatly swollen immediately before the siege because of a fair being held there, and because of refugees fleeing before the Persian army. The defenders suffered from disease during the siege but Ammianus does not indicate if food shortages became acute. Vegetius mentions the importance of building up supplies before a siege, but that would not have been possible in this case because of the speed of the Persian advance. Josephus frequently mentions the effects of food shortages on the Jews at Jerusalem: they were gradually starving and reduced to eating vermin or worse. Several historical accounts of sieges mention cases of cannibalism and Josephus gratuitously includes a story of a woman who allegedly killed and roasted her own baby. Appian accuses the Athenians of cannibalism during Sulla's siege in 87 BC, but Josephus in particular is prone to exaggeration, and we must be cautious in accepting as reliable anecdotes like this.

We have seen how Sertorius was able to smuggle skins of water into the garrison of Langobrigae, and Frontinus provides several stratagems on preventing supplies arriving, and obtaining supplies in time of blockade:

> When Hannibal was besieging Casilinum, the Romans sent grain in large jars down the river Volturnus, to be fished out by the besieged. When Hannibal prevented this by throwing a chain across the river, the Romans threw in nuts. Carried by the current, these floated down to the city, and with these provisions they provided the necessities for their allies.
>
> Frontinus, *Stratagems* 3.14.2

He also lists several examples of how the besieged might give the impression that they were well supplied to demoralize the enemy and make them think a long siege could be withstood, frequently by hurling supplies at the besiegers. Josephus, who reports several cunning stratagems that he devised in the defence of Jotapata, claims to have hung cloths soaked in water from the walls to convince the Romans that they had plenty of water. Conversely, the besiegers could attempt to demoralize the besieged by flaunting their own supplies as the Romans did to the starving Jews at Jerusalem.[177]

It was possible to aggravate the scarcity of food in a besieged town; whilst Vegetius advises that a town about to be besieged should send away all the non-combatants, Onasander suggests that the besieging general should send into to the town all prisoners except men of military age. These new arrivals would be useless in action but would consume the supplies more quickly. Caesar may have had this intention when he refused to allow the Mandubii through his siege lines at Alesia after they had been expelled from the *oppidum* by the Gauls precisely because of lack of supplies. Since the Gauls refused to allow them back in, the civilians were stuck in no-man's land between *oppidum* and siegeworks where they gradually starved. A similar incident occurred in the Roman siege of Cremna in AD 278; the rebels defending the town sent out the women and children because of food shortages but the Romans sent them back. Lydius, the rebel leader, then had them thrown into ravines around the town to conserve his food supplies.[178]

Sneaky tactics and stratagems

The principal aim of the besieging general would have been to capture the objective as quickly as possible and with the minumum of casualties to his own side. Whether this would be by assault or blockade would depend on particular circumstances, but military treatises provide plenty of advice and examples of ending sieges quickly, through either stealth or deception, or by encouraging surrender. Frontinus notes that Scipio captured New Carthage in 209 BC by attacking the walls across a shallow lagoon at low tide where the enemy were not expecting him, and both he and Sallust record how Marius captured a hill-town during the Jugurthine War through a small force which entered the town secretly. A Ligurian auxiliary looking for snails had found a way into the fortress up a cliff so steep that no sentries kept watch on it. The Ligurian led up a small force that included four centurions and the five most agile trumpeters and hornblowers, whilst Marius carried out a diversionary attack. When the Roman trumpets and horns sounded in the fortress, the defenders thought it had been captured and fled, allowing the main Roman force in. This ploy was also, according to Roman tradition, attempted by the Gauls with the Capitol in Rome in 390 BC, but failed because of some vigilant geese who alerted the Roman garrison. Frontinus also cites examples of pretended retreats by a besieging army, and of hostile forces gaining entry by disguising themselves as friendly forces, or even in one case as women, and then opening up the gates to the main force.[179]

Onasander advised a general to show mercy to a town that surrendered because in the course of a campaign this might encourage other towns to surrender, hopeful that they might be treated well too. During the course of a siege, however, it seems to have been just as likely that attempts would have been made to terrorize the besieged into surrender. Sulla is reported to have broken the resistance of the besieged at Praeneste during the Civil War of 82 BC by fastening on spears the heads of enemy leaders killed in battle and displaying them before the walls of the town. The besieged, demoralized by the deaths of their leaders, surrendered. Corbulo, who had captured Volandum in a speedy assault, caused the Armenian city of Tigranocerta to surrender by executing a captured Armenian nobleman and firing his head by catapult into the town. Frontinus claims that it landed in the middle of a council meeting whereupon the town immediately surrendered. The Roman commander besieging Machaerus at the end of the Jewish Revolt of AD 68-74 discovered that he had captured a popular youth called Eleazar. He made as if to crucify the youth, upon which the town surrendered to save his life, on agreement that Bassus would allow the defenders to go free, which he did.[180]

Surrender, sacking cities and the rules of war

Although the textbooks advise the general to encourage the city he is besieging to surrender, they do not suggest that the general make a formal request for a city to do this on his approach, and such requests are comparatively rare. Julian made such a request to the defenders of the Persian fortress of Anatha in AD 363 with the hope of avoiding a long and dangerous siege, but most requests of these sorts that are recorded in Roman histories come from non-Roman, particularly 'barbarian' enemies attacking Roman-held fortifications, and we have already seen that their abilities at siege warfare might be limited. Such armies might have had even more urgency than Roman armies to bring a siege to a swift conclusion. Titus requested the defenders at Jerusalem to surrender, but he had been carrying on siege works for some time under difficult circumstances and clearly wished to avoid further effort and losses. He made one particularly impassioned request to the Jews when he had failed to capture the strongest part of the city by assault and was about to resort to blockade. Despite advice in treatises about encouraging the defenders to surrender, it does seem to have been comparatively rare in Roman warfare in most periods for the Roman commander to request a surrender. Such an advance could be interpreted by the defenders as a sign of weakness of forces or of resolve, and it is not surprising that the Jews refused the requests they received, particularly since previous assaults had failed to capture the objective.[181]

Rules and conventions about surrender and the treatment of captured cities were not as well defined in the ancient world as they became in later periods. In the Medieval period, there were strict laws and conventions governing the conduct of siege warfare. On arrival at his objective, an attacking general was obliged to send a messenger ordering the defenders to surrender. Once the siege had begun (following a signal, such as a cannon shot), various rules applied concerning surrender, capture and the treatment of the inhabitants. Binding arrangements might even be made concerning the arrival of relieving armies and the surrender of a town if such a force failed to arrive within a certain time. Whilst some medieval conventions on siege warfare arose from Christian values, others had been in existence in the ancient world too, particularly the differing treatment of towns which surrendered and those taken by storm. In Roman warfare, the initiative in seeking surrender was expected to come from the besieged rather than the besieger. Caesar records a number of occasions when the inhabitants of Gallic *oppida*, which he was besieging or about to besiege, sent envoys to him to discuss surrender. Clearly terms might be discussed, as Josephus indicates happened at Machaerus, and Caesar himself at Alesia; in the former case the Jews agreed to surrender to the Roman commander provided that they were allowed to depart unharmed. At Hatra in the early third century, Septimius Severus paused for a day once his men had breached the outer of Hatra's two walls, expecting the defenders to come to terms voluntarily rather than be captured by assault and enslaved. When no peace overtures were forthcoming, he tried to continue the siege, unsuccessfully. By the fourth century AD, however, more formalized conventions concerning surrender seem to have developed, and perhaps these should be seen as part of an evolution towards the more rigid conventions of the medieval world. These are illustrated in the wars between Rome and Parthia and the civil wars recorded by Ammianus Marcellinus. In this period, the defenders of towns were more often invited to surrender at the beginning of a siege, and sometimes given a few days to consider the request, during which the enemy army would probably have begun to prepare siege equipment. As mentioned above, the fortress of Anatha surrendered to Julian, and in the Persian campaign of AD 359 two Roman forts, Reman and Busan, surrendered of their own volition to Sapor on his approach. In these cases the inhabitants were treated well, partly no doubt to encourage other places to surrender, as Onasander advised. The subsequent refusal of Amida to surrender to Sapor must have come as something of a surprise.[182]

There was a convention in ancient siege warfare that mercy should be shown to those who surrendered. This is not only advocated by the writers of military manuals, but also made clear by other writers. According to Tacitus, the Romans considered it barbaric to slaughter men who had surrendered (though this did not stop them from refusing to accept the surrender of an enemy and slaughtering them), and Livy records the general Aemilius Regillus refusing to allow his soldiers to sack the town of Phocaea in 190 BC after it had surrendered because 'cities were sacked after being taken by storm, not after surrender'. Although this convention existed partly to encourage surrender, it was not always followed, and when a violation did occur, Roman writers sometimes commented on it, or felt the need to provide an explanation. Sallust reports that the town of Capsa in North Africa surrendered to the Romans during the Jugurthine War, but even so the town was sacked, the defenders massacred and the rest of the population sold into slavery, a violation, he states, of the rules of war. He claims that Marius allowed the sack not because of cruelty or greed but because the people were untrustworthy. Not surprisingly, the episode greatly increased Marius' popularity with his own soldiers who had been given an unexpected opportunity to plunder. Like Capsa, Locha in Africa was sacked in 203 BC and the population massacred even though they surrendered as the Romans were assaulting it. The soldiers ignored the signal recalling them, and Appian explains that this was because of the hardships they had experienced during the siege.[183]

In Roman law, the general was entitled to treat the defenders and inhabitants of a besieged town in whatever way he liked once the siege had begun, but Roman attitudes towards the treatment of surrendered towns seem to have altered slightly during the later Republican period, with more willingness to be merciful towards the enemy in siege warfare. This is reflected most strongly in Cicero's albeit slightly idealistic suggestion that mercy should be extended to those who surrendered at whatever stage of the siege, and not just at the beginning, before the battering ram had been brought into action. This shift to a more merciful approach is probably linked with Rome's policy of expansion and the assimilation of new areas into her empire with the encouragement of co-operation.[184] It was usually only when that co-operation failed and rebellion occurred that these conventions were no longer in operation. Unless there existed specific reasons for the defenders and civilian population to be treated brutally, it was usual for those who surrendered to be spared. This did not necessarily spare them from slavery, and the population of the hill fortress of Numantia was sold into slavery in 133 BC after they were forced through starvation to surrender. But if the prospect of merciful treatment were to act as an encouragement to surrender, then the treatment of towns which did not surrender, and were taken by storm, was likely to be correspondingly harsh.

Josephus' description of the sack of Jerusalem illustrates the totality of siege warfare and the violence of a city taken by storm. After bursting into the Sanctuary area of the city, the Roman soldiers massacred indiscriminately. They killed not only men of military age and the elderly who would have had little monetary value as slaves, but also slaughtered indiscriminately women and children. This massacre, along with fires set by the Romans, shouting and looting helped to create an atmosphere of confusion and panic, but the soldiers had been given permission by their commander Titus to do what they wished to the city, the reward for their endeavours. Once the final area of Jerusalem, the Upper City, had been taken, again by assault, the Romans were able to sate their bloodlust and Josephus reports that they grew weary of killing. At this point, the historian reports that Titus ordered his men to kill only those who continued to offer armed resistance; all others were to be taken alive. Again, the sick and elderly were killed, and although

Josephus gives no reason for this exception, it is clear that it was because they were worthless as booty and therefore simply not worth sparing. Eventually the whole city, with the exception of three towers, was destroyed.[185]

Onasander recommended that only those bearing arms should be killed in the sack following the capture of a town by assault. Like many writers, he notes that desperate men with no hope of safety would only fight harder, and the purpose of the advice is to encourage the defenders to drop their weapons and offer no resistance. Clearly, this did not happen initially at Jerusalem and there was a general massacre of the population, as Josephus comments. After the capture of New Carthage by Scipio in 209 BC, the massacre was also indiscriminate, and Polybius notes that dogs and other animals were killed along with the humans. In both cases, however, parts of the city remained in enemy hands. At New Carthage, the Carthaginians still occupied the citadel, but once they realized the city was captured and surrendered themselves, Scipio called a halt to the general massacre. Though countering the advice of Onasander, the actions were nonetheless as effective in encouraging surrender as Onasander claims his advice was. The particular circumstances of each individual siege, the need to reward the soldiers and the strategic aims of the general must have influenced the treatment of cities taken by storm or which surrendered.[186]

The treatment of Marseilles by Caesar in 49 BC during the Civil War is an excellent example of this. The city had sided with Pompey and had caused Caesar's soldiers considerable difficulties and hardships during the long siege operations. The Massiliotes had also broken a cease-fire during which they had sortied and burned Caesar's siegeworks. Under such circumstances, when the defenders, short of food and weakened by disease, surrendered, it would not have been surprising had Caesar allowed his soldiers to sack the city. He did not, however. He claims that he spared Marseilles 'because of the famous name and antiquity of the city rather than because of any benefit it might bring to him'.[187] Caesar's very statement indicates that the value of a city might affect the way it was treated. Although advertising his famed clemency here, it is more likely that Caesar treated Marseilles mercifully because of its value as a military and naval base, and the local goodwill such a gesture would engender. He could also afford such a move strategically. The fall of Marseilles meant the end, for the time being, of serious resistance to Caesar in the north western Mediterranean, and he had no need to make an example of this city.

These conventions on siege warfare were less likely to be upheld in times of revolt and civil war, or if the enemy had committed atrocities. The massacre of Gauls at Avaricum in 52 BC was particularly savage; Caesar records that of the 40,000 inhabitants and defenders barely 800 escaped the slaughter when his soldiers took the *oppidum* by assault. Even the women and children were killed. The soldiers were not interested in plundering, Caesar claims, only killing, because of the difficulties of the siege, and because they wished to avenge the murder by the Gauls of Roman merchants at Cenabum a short time previously. Appian claims that Hasdrubal tortured and killed in full view of their colleagues Roman soldiers captured in the siege of Carthage in 146 BC specifically to destroy any chance of a surrender. Although Scipio Aemilianus did not take revenge on the general population of Carthage when he took the city by storm, Appian's statement indicates that the normal conventions could be ignored if atrocities had been committed. In the anarchy of civil war, all rules or accepted codes of behaviour might be ignored. Military discipline was generally lax, and prisoners were usually worthless since citizens could not be sold into slavery. As Tacitus' graphic and detailed description of the capture of Cremona in the civil war of AD 69 demonstrates, the sack could be particularly thorough. Cities that had revolted from Rome, however they were retaken, were also likely to be dealt with very harshly to serve as an example to others. When Capua surrendered unconditionally to the Romans in 211 BC, they considered destroying the town completely because the Capuans had defected to Hannibal and had tortured to death the Roman garrison and citizens who lived there. The town survived, but was heavily punished.[188]

It is perhaps not surprising that the defenders of Masada chose to commit suicide rather than face the Roman assault. The siege had been very hard and the defenders were rebels, according to Josephus of a particularly brutal disposition. Vespasian and Titus had already held their triumph for putting down the Jewish revolt so the Masada campaign was not of major military significance. Although Josephus has Eleazar, the leader of the Sicarii, claim they should avoid slavery by committing suicide, it is far more likely that they avoided being slaughtered, for an indiscriminate massacre of the defenders would probably have followed the Roman assault.[189]

Booty

> It is a law established for all time for all men that when a city is captured in war, the persons and possessions of all those in the city belong to the victors.
>
> Xenophon, *Cyropaedia* 7.5.73

We have seen how the treatment of captured or surrendered cities might vary with the circumstances of the siege and capture of the town, and also the value of merciful or harsh treatment of the town in the context of the campaign as a whole. Military service in the Republic was seen by all ranks as an opportunity for enrichment through the acquisition of booty, and pillaging would inevitably have taken place after the capture of a city, either officially sanctioned by the general or more surreptitiously.[190] Although during the period of the Republic the disposal of booty captured by the Romans seems to have been the prerogative of the commanding general, it was unusual for the soldiers not to be granted at least some of the captured plunder. Generals could make themselves very popular with the troops by allowing them the opportunity to enrich themselves in this way. The general himself was frequently a major beneficiary of the plunder of a captured town, but there were cases of generals being prosecuted by Tribunes of the Plebs if the soldiers received little or nothing. The state treasury or *aerarium* also frequently benefited from the sale of plunder, as happened at New Carthage. The quaestor, a Roman magistrate with the army who had responsibility for financial affairs, produced a record of the official plunder, and it may have been from such a source that Livy obtained the details of the spoils from New Carthage:

> 120 of the largest catapults, 281 smaller ones; 23 large *ballistae*, 52 smaller ones; large numbers of scorpions, both large and small, and weapons and missiles; 74 military standards. In addition, a large amount of gold and silver was brought to

> the general: 276 gold plates, almost all a pound in weight; 18,300 pounds of silver in coinage or ingots, and a large number of silver vessels. All this was weighed and counted for the Quaestor Gaius Flaminius. There were 40,000 modii of wheat, 270,000 of barley. 63 merchant ships in the harbour were attacked and captured; some with their cargoes of corn and weapons, along with bronze and iron, sail-cloth, esparto for making ropes, and other naval supplies for building a fleet.
>
> Livy 26.47

Polybius describes the sack and plundering of New Carthage by the Romans as being highly disciplined and organized, with only a proportion of the soldiers involved in these activities whilst the rest remained on alert. His description though is idealistic, and in reality the plundering of a city was no doubt haphazard and violent, involving rape, murder and opportunistic theft by the soldiers. Merchants accompanying the army on campaign could have played a valuable role in the selling off of the soldier's booty and exchanging it for more easily portable wealth in the form of coinage. In the Empire, the emperor or imperial treasury was likely to have been the principal beneficiary of spoils taken from a captured city. As Josephus indicates at Jerusalem though, the common soldiery had many opportunities to enrich themselves through plundering the city during the capture, and theft from Jewish refugees escaping from the besieged city. He reports, possibly somewhat dubiously, that soldiers in the Roman army took to cutting open refugees rumoured to have swallowed gold coins before escaping.[191]

Conclusions

> The general will use the many and varied types of siege engines as the opportunity arises. It is not for me to say that he should use battering rams, 'city-takers', 'sambucas' or mobile towers or covered sheds or catapults. All this depends on the luck, wealth and power of those fighting and on the skills of the architects accompanying the army to build the engines.
>
> Onasander 42.3

Onasander points out the number of variables in siege warfare, and the senselessness of going into much detail on how and when a besieging general should use pieces of equipment, but he does note the importance of wealth and resources in this type of warfare. In siege warfare, both attackers and defenders usually had time to react and devise methods of attack or defence; it was often a case of action and reaction by both sides. Josephus' defences at Jotapata, for example, were in reaction to the Romans' efforts: increasing the height of the wall against the Roman siege ramp; lowering chaff-filled sacks to cushion the blows of the battering ram (which the Romans then cut down with reaping hooks attached to long poles); pouring boiling oil on the Roman *testudo* formation; and pouring boiled fenugreek on the gangway boards. Equipment and supplies could be prepared, as Vegetius advised, but Onasander is right when he says that the general must use the available equipment as the opportunity arises.

57 *Titus processing through Rome with booty from the sack of Jerusalem*

Siege warfare changed little from the Assyrian period to the late Roman. The basic equipment was essentially the same, as were the methods of countering it. Assyrian reliefs show techniques of siege warfare, including defenders catching the heads of rams in nooses, which Vegetius recommends over a thousand years later. The major changes were the invention of torsion artillery in the fourth century BC and the use of circumvallations, neither of which was a Roman innovation. Only refinements to these and to the conventions on siege warfare were made by the Romans.

Although the enthusiasm with which they prosecuted a siege might vary, the Romans virtually never abandoned a siege, once begun. Scipio's predecessors at Numantia had conducted what seems to have been at times a rather half-hearted blockade or even just close observation of the fortified hill-town, but eventually, with the construction of a circumvallation, the blockade was enforced and the population forced to surrender. To have abandoned such a task, once begun, would have set a very dangerous example for a power which, like Rome in that period, was aggressively expansionist. Rome's policies of expansion and assimilation of conquered territory into provinces also influenced the way captured or surrendered towns were treated, but the 'standard' conventions on the treatment of the defeated enemy in siege warfare were generally accepted throughout the ancient world. Because of its militaristic nature, however, Rome possessed the resources, the logistical organization, the ability to keep an army in the field for a long time, and the skills necessary for conducting highly successful siege warfare.

Conclusions

One of the most prominent features of the Roman army that has emerged from this study is its enormous flexibility. In its organization, use of equipment and fighting technique, the Roman army could adapt itself to cope with the different types of enemy, fighting and terrain that it encountered. Contemporary authors boasted that one of the reasons for its success was that the Roman army was open to change: the Romans were prepared to adopt equipment and fighting techniques from their enemies and adapt them for their own use. Tradition has the Romans adopting the close infantry fighting with *scutum* from the Samnites, and the *gladius Hispaniensis* from the Spanish tribes. It was not so much a case of beating the enemy at his own game, as changing the rules so that Rome won. We find innovation in use of weapons in Italy in the third century BC to deal with the long but bendy swords of the Gauls; in Spain in the second century BC, in the form of cohorts, to cope with the local fighting styles and the terrain; and in Cappadocia in the second century AD to confront the mounted threat of the Alans. The ability to adapt comparatively easily to different enemies, equipment, terrain and circumstances was enhanced by the Roman attention to training and discipline. Even during the Republic, when Rome did not retain a standing army, many generals saw these matters as being of considerable importance. Scipio Aemilianus re-imposed discipline on his troops in Spain and sent them through a rigorous training programme before commencing his campaign at Numantia. Metellus similarly paused before continuing the war against Jugurtha, and Marius did the same in some of his campaigns. He trained his legions for months before allowing them to face the Cimbri and Teutones in battle, as well as hardening them and making them eager for combat through making them work on civil engineering projects. In the imperial period, the existence of a 'professional' standing army should have allowed for the constant maintenance of a high level of discipline and training. This should have allowed the soldiers to adapt fairly easily to new equipment and drills, such as those demanded by Arrian's proposed tactics against the Alans. Even so, we still find disciplinarians of the Republican school like Corbulo in the East who kept his men under canvass throughout the hard Armenian winter to toughen them up. Frontinus also reports the punishment of an auxiliary commander who had kept his *ala* of cavalry inadequately equipped. These stories may have been reported because Corbulo was exceptional in the standards of discipline he required, and indeed Tacitus indicates that this was the case. Other provincial governors and generals may have been much slacker in the standards of training and discipline they required.

We saw how Trajan made his men train on their long march east for the Parthian campaigns, no doubt learning how to deal with the different types of enemy, warfare and terrain that they would experience. There was undoubtedly variation of practice in the Roman army in different provinces and areas of the empire, but we have seen how armies could cope with these, particularly well trained veteran armies like Caesar's in the Civil War. Though experiencing difficulties initially in Africa and Spain because of the local fighting fashions being used by their opponents, Caesar's soldiers quickly adapted and were able to cope with these differences. Soldiers stationed in one area, used to fighting against a particular enemy, tended to take on some of the customs and characteristics of enemy fighting practices. We should therefore not be surprised to find regional differences in Roman military practices, particularly in the late Republic and Empire when units were stationed for long periods in the same place. The equipment and tactics proposed by Arrian against the Alan cavalry would not have been so useful against the infantry and lighter cavalry that Roman armies were facing in other parts of the empire.

There existed in the Imperial period a set of military regulations or *constitutiones* issued by emperors, that covered training and may have dealt with other aspects of Roman campaigning. We learn from Tacitus of two military regulations, prohibitions on falling out on the march and on fighting without orders, re-imposed by Corbulo on his army when he was governor of Germany. We have already noted how the 'good' general was expected to ride up and down the marching column to prevent men from falling out without permission. Frontinus' *Stratagems* indicate that the second regulation at least was a very old one indeed: T.Manlius Torquatus had executed his son in 340 BC for engaging in battle with the enemy against orders, even though he was victorious. Other possible regulations are reported by Frontinus in his account of the military oath sworn by soldiers. This included a pledge not to flee from battle, and not to break ranks unless to find a weapon, strike the enemy, or save a comrade. We do not know, however, the range of these regulations, how official they were, or the extent to which they were imposed. Corbulo clearly had a reputation for being unfashionably strict when it came to discipline and regulations, but it was probably up to the provincial governor or army commander to set the tone. Scipio Aemilianus may have criticised one of his men for an elaborately decorated shield, and smashed up the decorative utensils of his soldiers, but Caesar did not care how richly embossed his soldier's equipment was.

It is Caesar who suggests that there were standard ways of doing things in the field, when he says that his deployments against the Nervii were dictated more by the terrain and the immediacy of the situation than the demands of military theory. As Onasander admits, theory can only cover a limited number of options; it cannot cope with the unexpected. When the unexpected happens, a general might employ a stratagem of the type collected by Frontinus. His *Stratagems* were intended, he claims, not for generals to copy blindly, but to inspire them to devise their own, and give them confidence from the successful plans of their predecessors. They too have a place in the education of generals according to Polybius, and the examples certainly illustrate well much of the advice provided in the textbooks, justifying the suggestion that Frontinus intended his *Stratagems* to be an appendix to his own general treatise on the art of war.

A textbook that accepts its limitations is more valuable than one that tries to cover every eventuality: Vegetius is very prescriptive, particularly on battle dispositions, and he cannot cover all variations and possibilities. Onasander knows that war involves too many variables for him to be comprehensive in his advice, and there is always the random factor of luck, or *fortuna*, which could ruin even the best laid plans. It is not surprising that he lists luck as being one of the qualities necessary for a good general, and that many of the great generals of the late Republic claimed to be favoured by *fortuna*. Sulla took the name Felix to boast his own luck, Sertorius kept a white hart, which brought him luck, and Caesar stressed the role of *fortuna* in his victories (but also blamed it for his defeats). Because of the variables in war, Onasander only gives general advice on deployment for pitched battle, and leaves it up to the general to decide how best to use his various items of siege equipment, because that will depend on the circumstances of the particular siege.

As a means of learning the arts of generalship and warfare, the textbooks can provide a realistic and practical source of information. Cicero might have expressed disdain about some of his contemporaries who learned the art of generalship through books (even though he probably did himself), but Polybius was being genuine when he claimed that one could learn from such books. The general treatises that have survived, particularly those by Onasander and Vegetius, do provide a great deal of valuable information concerning field practices. As we have seen, that information, though often basic, is nonetheless valuable and practical because it is reflected in the actual field practices of the Roman army as recounted by historians and revealed by archaeology. A mediocre general might have been able to get by with this essential knowledge and the advice and experience of his officers; a great general, as Onasander implies and Caesar was aware, would know when to move beyond the recommendations of textbooks and trust to his own imagination and inspiration.

Notes

1. Cicero's praise of Lucullus, *Lucullus* 1.1-2; and of Marius and Pompey, *Pro Balbo* 47, *de Imperio* 27-8.
2. For a discussion of Roman politics and the causes of expansion, see W.V.Harris, *War and Imperialism in Republican Rome 327-70 BC*, 1985; Cicero describes briefly his victories in Cilicia in *Letters to Atticus* 5.20.
3. Sallust, *B.Jug.* 37-39 on Spurius Albinus. Although Aulus Albinus agreed terms with Jugurtha to preserve his army, Rome was immediately able to repudiate the treaty because it had not been ratified by the Senate and People (by which time the troops had been removed from danger); for Domitius' appointment, see Plutarch, *Cicero* 37. Cicero allegedly retorted that the man was better qualified to be a school teacher.
4. eg: Webster, *The Roman Imperial Army*, 1985, 113-6.
5. Pliny the Younger, *Letters* 3.11; Tacitus, *Agricola* 3.
6. On the qualities of generals, see Cicero, *de Imperio* 28; 36 and Onasander 1.2. Polybius lists his three sources of knowledge at 11.8.1. Cicero on the use of handbooks, *Pro Fonteio* 42, *Pro Balbo* 47, *Lucullus* 1.
7. See C.M.Gilliver, 'Mons Graupius and the role of auxiliaries in battle', *Greece & Rome* xliii.1 (1996), 54-67.
8. Livy's description of the legion: 8.8; Keppie discusses the effect of the Gallic threat on Roman military organization (*The Making of the Roman Army*, 1984, 19). Diodorus Siculus 23.2, Athenaeus 273 F and Sallust *Cat.*55.38 all include the literary tradition or *topos* of Romans adopting the weaponry and tactics of their enemies, then turning the tables on them. Arrian lists several 'barbarian' practices that Hadrian introduced into cavalry exercises, 44.1, and the tenth century lexicon *Suda* notes that the Romans adopted the *gladius Hispaniensis* from the Spanish tribes (303.1).
9. Varro, *de Lingua Latina* 7.58.
10. For a full discussion of legion strengths, see Brunt, *Italian Manpower*, 1970, appendices 25 & 27.
11. Cincius Alimentus is cited by Aulus Gellius, *Attic Nights* 16.4.6; Parker (*The Roman Legions*, 1928, 28) and Speidel (*The Framework of an Imperial Legion*, 1992, 7) are among the historians who attribute the introduction of the cohortal legion to Marius. For Marius' introduction of the eagle as the principal legionary emblem, see Pliny, *Natural History*, 10.5.
12. See M.J.V.Bell, 'Tactical Reform in the Roman Republican Army', *Historia* xiv (1965) 404-422. Bell discusses all the evidence for cohorts and presents a compelling argument for their use in this particular theatre of war.
13. Sallust, *B.Jug.* 46.7 and 50.1 refer to cohorts of light-armed Roman legionaries (*expeditae cohortes*), but *B.Jug.* 100.3 to light-armed maniples. *B.Jug.* 49.2 describes the Numidian king Jugurtha going round the 'maniples' of his army.
14. Pseudo-Hyginus 3 and Vegetius 2.6 on the milliary first cohort.
15. Vegetius 2.12 on the 'cohort commander'. A.K.Goldsworthy argues that the senior centurion commanded the cohort, *The Roman Army at War*, 1996, 15.
16. D.B.Saddington provides the best discussion of the development of these auxiliary units in *The Development of the Roman Auxiliary Forces from Caesar to Vespasian*, 1982.
17. A *turma* may have contained either 30 or 32 men; the evidence is unclear.
18. For a brief discussion of the papyrological evidence for unit size, see M.W.C.Hassall, 'The Internal Planning of Roman Auxiliary Forts' in ed. B.Hartley & J.Wacher, *Rome and her Northern Provinces*, 1983, 96-131; for the strength report or *pridianum* of Cohors I Tungrorum from Vindolanda see A.K.Bowman & J.D.Thomas, 'A Military Strength Report from Vindolanda', *JRS* 81 (1991) 62-73.
19. Tacitus, *Annals* 4.47; 14.23; Arrian, *ektaxis* 7. Vespasian's army in the Jewish war included contingents provided by client and pro-Roman kings, including an Arabian ruler Malchus, Josephus, *Jewish War* 3.68.

20 Jarrett most recently argued in favour of 'legionary commands' in 'Non-legionary Troops in Roman Britain: Part one, The Units', *Britannia* 25 (1994) 35-77; M.Roxan has opposed this view in 'Roman Military Diplomata and Topography', in *Studien zu den Militärgrenzen Roms III*, 1986, 768-778. For Tacitus on the Batavians, see *Histories* 1.59.
21 Vegetius 1.1; Onasander 6; 10; Frontinus *passim*; Goldsworthy, 1996 and Ferrill, *The Fall of the Roman Empire: the Military Explanation*, 1986, consider in detail Roman military discipline.
22 Vegetius 3.1.
23 The manpower figures are discussed in detail by P.A.Brunt, *Italian Manpower*, 1971, Appendix 26. For Velleius Paterculus' claim, see 2.15.
24 Diodorus 36.1 and Livy *Per.* 67 give the figures for Arausio; for Mithridates, Plutarch, *Sulla* 15 and Appian, *Mithridatic Wars* 45; the Boudiccan revolt, Tacitus, *Annals* 14.34; Dio 62.8.2.
25 Tacitus *Annals* 2.16.
26 Appian, *Spanish Wars* 85; Sallust, *B.Jug.*45; Plutarch, *Marius* 13.
27 Tacitus, *Histories* 2.87; 3.33; M.P.Speidel suggests 500 servants to each legion, 'The Soldiers' Servants', *MAVORS* 8, 1992, 342-350.
28 Vegetius 3.6; Caesar *B.G.* 6.40.
29 Servants joining in the war cry, Frontinus, *Strat.*2.4.8; Livy 7.14; as 'reinforcements', *Strat.*2.4.1; Livy 7.14.
30 For the Punic War, see Livy 27.18; the battle against the Nervii, Caesar, *B.G.* 2.27.
31 The massacre at Cenabum, Caesar, *B.G.* 7.3; Quintus Cicero's camp, Caesar, *B.G.* 6.37.
32 Trasimene, Livy 22.4-7 (here the army was advancing); Varus, Suetonius, *Augustus* 23. Other particularly severe defeats were encountered by Sabinus and Cotta in Gaul in the winter of 54-53 BC (Caesar *BG* 5.31-37) and by Cestius in Judaea in AD 67 (Josephus 2.540). Both armies, like that of Varus, were in retreat. For Dio's account of the Varian disaster, 56.20-21. Velleius Paterculus, 2.119, has a more concise account of the episode, concentrating more on the cowardly behaviour of the officers than the dramatic details preferred by Dio.
33 *Kalkriese – Römer im Osnabrücker Land*, ed. W.Schlüter, 1993.
34 Vegetius 2.27 for the *constitutiones* of Augustus and Hadrian; 1.9, 19 & 27 for practice marches; Onasander 6; Tacitus *Annals* 11.18 for regulations on marching columns. Corbulo, renowned as a disciplinarian, reintroduced this regulation when governor of Germania Inferior in AD 47 along with another regulation, on not fighting without orders.
35 Sallust, *B.Jug.* 45 for Metellus; 96, Sulla; 100, Marius; Tacitus, *Agric.* 20 on Agricola; *Histories* 2.5 for Vespasian; Dio 68.23 on Trajan; Arrian, *ektaxis* 10 on his own intentions and Caesar, *B.G.* 7.67 on his own actions. Caesar *B.G.* 2.21 on the importance of standards in deployment.
36 Tacitus *Histories* 2.89; Josephus *B.Jud* 3.115ff; 5.39ff. Josephus' descriptions, particularly the more detailed one of Vespasian's initial march into Galilee, add to his portrayal of an invincible army and the futility of Jewish resistance. This, and his fascination for all things relating to the Roman army, help to explain why he goes into details that are rarely found in other authors. Lucian, *How to Write History* 37 & 39.
37 Caesar, *B.G.* 2.18-19.
38 Sallust, *B.Jug.* 46; 100 on the marching formations in Africa; on the column ready for battle, Seneca, *Ep.* 59.7.3. The formation is termed the *agmen quadratum* or squared column.
39 F.Walbank, *An Historical Commentary on Polybius*, 1957, 723.
40 Polybius 6.40 and Josephus 3.97 on the order of legions within the marching column.
41 On the use of scouts, Arrian, *ektaxis* 11; Caesar, *B.G.* 1.41; see also Austin & Rankov, *Exploratio*, 1995. On engineers and road-builders, Josephus, *B.Jud.* 3.141; Pseudo-Hyginus 30.
42 Le Bohec, *The Imperial Roman Army*, 1993, 128-130 on marching formations; Anonymous Byzantine Stratagemata 18 on negotiating confined areas.
43 Caesar, *B.G.* 5.33.
44 Livy 39.30 on the *testudo*; see also Chapter 5 on siege warfare for more details of this formation. The aggressive nature of the *agmen quadratum* is shown in a simile of Cicero. He describes his great enemy Mark Antony entering a meeting of the Senate 'in an *agmen quadratum*', literally bristling hostility and domination(*Phillipics* 18.20). For Metellus in Africa, Sallust, *B.Jug.* 49. Metellus altered his marching formation to take account of the terrain and the proximity of the enemy.
45 Tacitus, *Histories* 2.68 on the panic in Vitellius' army; and on the dangers of a long column, *Annals* 2.5; Caesar, *B.G.* 4.31-33 for Sabinus and Cotta.

46 M.Gichon, 'Aspects of a Roman army in war according to the *Bellum Judaicum* of Josephus', in ed. Freeman & Kennedy, *The Defence of the Roman and Byzantine East*, 1986, 287-310.
47 Four part-mounted cohorts (III Ulpia Petraeorum, IV Raetorum, I Ituraeorum and I Germanorum) had cavalry present but no infantry; Legion XII was present only as a vexillation; and Cohors I Apulorum had only 200 men present, though it is possible that this was the entire unit.
48 Caesar's interception of the Aedui, *B.G.* 7.40-41, and rescue of Q.Cicero, 5.46-47. Gaius Claudius Nero marched 6,000 infantry and 1,000 cavalry some 250 miles from southern Italy to Picenum to support his consular colleague M.Livius Salinator in 207 BC. They completed the march in seven days and this speedy reinforcement of Salinator's army allowed the Romans to defeat Hasdrubal at the Metaurus, the first major Roman victory of the Second Punic War, Livy 27.43-49.
49 *B.Afr.* 69-75. A small force of Gallic tribesmen caused considerable difficulties to the progress of Hannibal's army by attacking it in a confined space in an Alpine pass, Livy 21.32-33.
50 Antony in Parthia, Dio 49.29; Caesar, *B.G.* 7.67; 5.33-34. Lack of discipline also contributed to the Varian disaster.
51 Trasimene, Livy 22.4-7 & Polybius 3.83-4; Lucius Manlius, Livy 22.25; Lucius Postumius, Livy 23.24.
52 Scipio, Livy, 25.34; and Cynoscephalae, Livy 33.7-10; Polybius, 18.21-27.
53 Frontinus, *Strat.* 2.1.14 for Lucullus' attack on Mithridates and Sallust, *B.Jug.* 49 for Metellus' alterations to his marching formation.
54 Livy 44.5 on the campaign in the Cambunian mountains of northern Greece. Such was the nature of the terrain that the army was reduced to travelling only seven miles on one day.
55 Josephus *B.Jud.* 3.55 on the soldier's kit. For a discussion of the military kit and its weight, see J.Roth, *The Logistics of the Roman Army at War*, 1999, 71-76.
56 Metellus Numidicus, Sallust, *B.Jug.* 76; his son, Metellus Pius, Plutarch, *Sertorius* 13; Cestius Gallus, Josephus, *B.Jud.* 2.545-6.
57 Vegetius 3.3 and Onasander 6 on supplies; papyrological evidence, C.E.Adams, 'Supplying the Roman Army: O.Petr.245', *ZPE* 109 (1995) 119-124. Appian, *Mithridatic War* 12 for Lucullus; Tacitus, *Annals* 2.5 for Germanicus' campaigns.
58 Vaga, Sallust, *B.Jug.* 47; the Pompeians, Appian, *B.Civ.* 2.95; Sallust's logistical role, *B.Afr.* 8; 34. On South Shields, Bidwell & Speak, *Excavations at South Shields Roman Fort, vol.1*, 1994. Other supply bases in Britain have been identified at Fishbourne dating to the early Claudian or Neronian period, and connected with the invasion and early conquests; and at Llanfor in Gwynnedd, dating to the late Claudian or Neronian period, and probably connected with advances into Wales.
59 Onasander 6 on foraging. Britain, Caesar, *B.G.*4.32; Sertorius, Frontinus, *Strat.* 2.5.31; Scipio Aemilianus, Appian, *Punica* 14.100. The story may not be true, but the concerns are genuine. Q.Cicero's foraging party was also attacked on its way back to camp in Gaul in 53 BC. He had allowed five cohorts of legionaries, and servants to go out and forage despite Caesar's orders that he keep them in the security of the camp. They were caught out in the open by German cavalry and suffered serious losses before some of them managed to push their way back into camp and Caesar's timely arrival put an end to the panic, Caesar, *B.G.* 6.36-42.
60 Corbulo in Armenia, Tacitus, *Annals* 14.23-26; the capture of Gomphi, Caesar, *B.Civ.* 3.80-1.
61 Dio 68.23.
62 Polybius 6.41 on Roman camps; Josephus *B.Jud.* 3.86 and Livy 44.39 both stress the routine of fortifying camp before a pitched battle.
63 Livy 22.49-52 on the aftermath of Trasimene.
64 Appian, *B.Civ.* 2.74-5. In his own account, Caesar does not claim to have 'burnt his boats' in the way that Appian does, but simply reports that he had destroyed his camp before a day's march, then seized the unexpected opportunity of battle with Pompey. Nonetheless, Appian's report illustrates well the importance of marching camps in pitched battles.
65 Tacitus, *Annals* 4.25 on Tacfarinus; the attack on Sabinus' camp, Caesar, *B.G.* 3.17-19. German tribes under Arminius attacked Germanicus' encampment the night before the battle at Idistaviso in AD 16. However, the attack was expected and the Germans found the defences lined with troops (Tacitus, *Annals* 2.13).
66 Frontinus provides several examples of this punishment, *Strat.* 4.1.18-19, 21; the last example is of Corbulo, the renowned disciplinarian.

67 Josephus, *B.Jud.* 3.86 and Vegetius 1.21 on the speed of construction; *Pseudo-Hyginus* 52 on the use of pickets.
68 On Corbulo, Frontinus, *Strat.* 4.7.2; on Pyrrhus, *Strat.* 4.1.14, Livy 35.14 and Plutarch, *Pyrrhus* 16.
69 Herodotus 9.15; 70 on the Persian camp at Plataea.
70 Aeneas Tacticus refers to his treatise on castrametation in his surviving work on siege warfare, mentioning the sections on guards and patrols (21.2).
71 Cicero, *Pro Rabirio* 42 (Caesar); Tacitus *Histories* 2.5 (Vespasian); Tacitus, *Agricola* 20 (Agricola); SHA *Hadrian* 10.6 (Hadrian).
72 This advice is collected from Polybius 6.27; Onasander 8; Pseudo-Hyginus 56-7 and Vegetius 1.22; 3.2 & 8. Much of the advice is common to all sources.
73 For *bracchia*, Caesar, *B.Civ.* 1.73; *B.Hisp.* 13; Plutarch, *Marius* 18.
74 Caesar, *B.Civ.* 1.48 for the camp in Spain; Tacitus, *Histories* 5.24 for the Batavian revolt.
75 Caesar, *B.G.* 3.28; Trajan's Column scenes 46-8; 138-140.
76 The gully at Swine Hill may have contributed to the defences on that side of the camp which were at the top of the steep slope. This advantage may have countered any concerns about the enemy sneaking up unseen.
77 Livy 44.3 on the overlooked camp.
78 Sallust, *B.Jug* 45 for Metellus in Africa; Appian, *B.Civ.* 3.83 on Antony and Lepidus. The troops of the two sides visited each other's camps. On Caesar, Appian, *B.Civ.* 1.74; the Pompeian leaders executed Caesar's soldiers who were found in their camp, but Caesar sent all his enemy's soldiers back safely. Actions like this, and display of the general's *clementia* or mercy encouraged desertions to Caesar's side. On Cerialis, Tacitus, *Histories* 4.75.
79 Pseudo-Hyginus 48; 51-53, Vegetius 1.24; 3.8 and Josephus, *B.Jud.* 3.84 on the size of camp defences.
80 Caesar, *B.G.* 2.5 and *B.Civ.* 1.41 on the dimensions of his camp defences.
81 Vegetius 1.24; 3.8 on ditches.
82 Pseudo-Hyginus 50-53 and Vegetius 1.24 on size of ramparts; Vegetius 3.10 on Persian practices.
83 Livy *Per.*57 & Dio 17.63 for Numantia; Caesar, *B.Civ.* 1.42 for encamping in Spain.
84 Vegetius 3.8 on *tribuli*; for a full discussion of this issue, see C.M.Gilliver, 'Hedgehogs, Caltrops and Palisade Stakes', *Journal of Military Equipment Studies* 4 (1993) 49-54.
85 On the location of 'Stracathro' type camps, see Hanson, *Agricola and the Conquest of the North*, 1987, 123-125.
86 For surveyors and an advance party, Polybius 6.41. Attacks on armies engaged in entrenching camp, Caesar, *B.G.* 2.19 and 3.28; *B.Afr.* 74; Appian, *B.Civ.* 5.110. The soldiers shown entrenching on Trajan's Column have piled their arms and armour neatly nearby in case of attack.
87 Infantry screens for entrenching armies, Livy 44.37; Caesar, *B.G.* 1.49; *B.Civ.* 1.41; for practice camps, see R.W.Davies, 'Roman Wales and Roman Military Practice-Camps', *Archaeologica Cambrensis* 117 (1968) 103-20.
88 For camp size, Pseudo-Hyginus 21, Vegetius 1.22; 3.8, and Onasander 10.16. Tracing campaigns through marching camps: eg: St. Joseph, 'Air Reconnaissance in Roman Britain, 1969-72', *JRS* 63 (1973) 214-246; 'The Camp at Durno, Aberdeenshire, and the site of Mons Graupius', *Britannia* 9 (1983) 271-87; and W.S.Hanson, *Agricola and the Conquest of the North*, 1987, 132-3.
89 For a discussion of some factors which might influence camp size, see W.S.Hanson, 'Roman Camps north of the Forth-Clyde isthmus: the evidence of the termporary camps', *PSAS* 109 (1978) 140-150. On deceiving the enemy, Onasander 10.16; Frontinus *Strat.* 1.1.9. Livy tells the story rather differently, reporting that Hasdrubal saw through the ruse, 27.43 ff.
90 St Joseph, 1978, 'The camp at Durno, Aberdeenshire, and the site of Mons Graupius', *Britannia* 9, 271-87.
91 Tacitus, *Agricola* 25. Caesar used the Saône to supply his army at the start of his campaign against the Helvetii, *B.G.* 1.16, and Germanicus transported both troops and supplies by sea and river for his operations against the Germans, Tacitus, *Annals* 1.70; 2.5.
92 Dio 62.1-12 on the Boudiccan revolt; Tacitus, *Agricola* 29-38 on Mons Graupius.
93 On training, Onasander 10.1; Josephus, *B.Jud.*3.75. Scipio Aemilianus (Appian, *Spanish Wars* 85), Metellus (Sallust, *B.Jug. 45*) and Corbulo (Tacitus, *Annals* 13.35) are all reported to have reinforced discipline in this way.
94 Vegetius 3.10 & 12. On the necessity of climbing over bodies of the dead on the battlefield, Josephus, *B.Jud.* 6.12.

95 Sallust, *B.Jug.*86 on Marius; Caesar, *B.G.*2.20 for the attack by the Nervii.
96 Onasander, 14.2, and Frontinus' examples, 1.11, especially 17-18; Plutarch, *Marius* 14-18.
97 Advice and 'stratagems' on the timing of battle, Vegetius 3.11; Onasander 12; Frontinus, *Strat.* 2.1; for Magnesia, *Strat.* 4.7.30 and Livy 37.41; Tacitus, *Histories* 3.19-22 for the battle of Cremona.
98 See Goodman & Holladay, 'Religious Scruples in Ancient Warfare', *Classical Quarterly* 36 (I), (1986), 151-171. Frontinus, *Strat.* 2.1.16 on Vespasian in Judaea; for Ariovistus, Caesar, *B.G.* 1.50; Frontinus, *Strat.* 2.1.17.
99 Frontinus, *Strat.* 2.1.13.
100 *B.Hisp*.28-30 for the battle of Munda; Polybius 2.33 for battle against the Insubrian Gauls.
101 Vegetius, 3.13; Mons Graupius, Tacitus, *Agricola* 35-37; Zela, *B.Alex.*75. During the civil wars in Spain the Pompeian commander Marcellus refused to let his army engage that of Q.Cassius because the latter had drawn up his troops on high ground and Marcellus knew his troops would be seriously disadvantaged if they did attempt to storm the hill, *B.Alex.* 60; cf. also Ostorius Scapula's troops successfully storming a defensive position in Wales, Tacitus, *Annals* 12.35.
102 For a more detailed discussion of the role of Batavians, see M.W.C.Hassall, 'Batavians and the Roman conquest of Britain', *Britannia* I (1970) 131-136; Caesar's battle at the Aisne, *B.G.* 2.8 9; Germanicus against the Cherusci, Tacitus, *Annals* 1.63-68.
103 The Batavian revolt, Tacitus, *Histories* 5.14-18.
104 Hannibal's dispositions at Cannae, Livy 22.43; Frontinus, *Strat.* 2.2.7; Polybius 3.14; Marius against the Cimbri and Teutones, Frontinus, *Strat.* 2.2.8; Plutarch, *Marius* 26. On the second battle of Cremona, Tacitus, *Histories* 3.23.
105 For Cynoscephalae, Polybius 28.26; and the flexibility of the cohort formation, Caesar, *B.Afr.*12 where Caesar ordered every other cohort to turn round to face a cavalry attack from all sides; Onasander 21 and Vegetius 3.20 on using ground to prevent a flank attack. For Magnesia, Appian, *Syrian Wars* 31, and Pharsalus, Caesar, *B.Civ.* 3.88-9. Suetonius Paulinus' dispositions against Boudicca, Tacitus, *Annals* 14.34; Arrian, *ektaxis* 11 for his proposals.
106 Physical obstacles on the battlefield: Frontinus, *Strat.* 2.3.17 for Sulla; Caesar, *B.G.*2.9 against the Belgae, and *B.Afr.*51-60 at Uzitta. The Bosporan king Pharnaces used trenches to protect his infantry against Domitius Calvinus in 48 BC, *B.Alex.*37. For Munda, see *B.Hisp.*30.
107 On the impression given by armour, Onasander 28; Cato at Pydna, Frontinus, *Strat.* 4.5.17, and 4.1.4 for the military oath.
108 Onasander 10.10 on the importance of the omens.
109 Frontinus, *Strat.* 1.12.2 on Scipio; for the eclipse at Pydna, see Livy 44.37 and Frontinus, *Strat.* 1.12.8. A similar story is told of William the Conqueror tripping on the beach on landing in Britain in 1066, but this may be a later addition to the tale, based on Roman examples. For a full discussion of military defeat and religious failings, see N.S.Rosenstein, *Imperatores Victi: Military Defeat and Aristocratic Competition in the Middle and Late Republic*, 1990, Chapter 2, 'Defeat and the *Pax Deorum*'.
110 On oratorical skills, Onasander 1; Pharsalus, Caesar, *B.Civ.*3.90. Caesar's list of the things to do before battle, and the attack by the Nervii are at *B.G.*2.20-21. Interestingly, the author of the *Bellum Alexandrinum* presents a variation of these actions when Caesar has to respond quickly when once again his men were caught entrenching camp, this time at Zela in 47 BC, *B.Alex.* 75.
111 On speeches, see 'The Battle Exhortation in Ancient Historiography. Fact or Fiction?' M.H.Hansen, *Historia* 42.2 (1993) 161-180. Hadrian's *adlocutio* is most easily accessible in translation in B.Campbell, *The Roman Army: a Sourcebook*, 1994, no.17.
112 Frontinus, *Strat.* 2.3. Frontinus gives 24 examples of dispositions, the other extensive selections relate to 'escaping from difficult situations', 1.5; ambushes, 2.5; discipline, 4.1; and 'sundry maxims and plans', 4.7. On war councils, see Caesar, *B.G.*4.13; 6.5; 7.45.
113 Lucullus, Appian, *Mithr.* 84-5; Zela, *B.Alex.* 75; Caesar at Pharsalus, *B.Civ.* 3.89 and Frontinus, *Strat.* 2.3.22-3; Scipio in Spain, Livy 28.14-15 and Frontinus, *Strat.* 2.3.4.
114 Vegetius 2.15 explains the deployment of the legion, but he is confused and elsewhere talks of *hastati, principes* and *triarii* in the lines. Most likely he is using two or more different sources dating to both the Republic and the Imperial period (including Cato the Elder, and possibly Frontinus for the later period), so his battle line is most probably a synthesis of arrangements from different periods. For Uzitta, see *B.Afr.* 81-84.
115 Caesar's battles against the Helvetii, *B.G.*1.25-26, and Ariovistus, *B.G.*1.52; Cynoscephalae,

Livy 33.9; Polybius' discussion of legion against phalanx, 18.30-32.

116 Auxiliaries at Pharsalus, Appian, *B.Civ.*2.75; Crassus in Aquitania, *B.G.*3.24; battle against Pharnaces, *B.Alex.*38-40. Deiotarus' legions were later combined and regularized as Legion XXII Deiotariana; Caesar's use of inexperienced troops, *B.G.*1.24; 1.50.

117 Uzitta, *B.Afr.*59-60; Pharsalus, Caesar, *B.Civ.*3.88; slingers and archers in the front ranks, *B.Afr.*12; 79, *cf.* Frontinus *Strat.* 4.7.27 on Scipio Aemilianus who distributed slingers and archers amongst all his legionaries at Numantia in 133 BC.

118 Issus, Dio 75; Onasander 20 on how heavy infantry should deal with light armed missile troops.

119 P.A.Holder, *The Auxilia from Augustus to Trajan*, 1980, appendix 3 lists all attested auxiliary units. Of the 32 units of archers, 4 were *alae*, 6 part-mounted; 4 were milliary, of which two were part mounted; Vegetius1.16 on training in the use sling and bow. Idistaviso, Tacitus, *Annals* 2.16.

120 Frontinus on Pharsalus, *Strat.*4.7.32; Caesar, *B.Civ.* 3.93-4.

121 The infantry and cavalry of mixed units separated on the march, Josephus, *B.Jud.* 3.125-26; Arrian, *ektaxis* 1.

122 Munda, *B.Hisp.* 30-31; Ariovistus, Caesar, *B.G.* 1.52; Mons Graupius, Tacitus, *Agricola* 37.

123 Suetonius Paulinus' dispositions, Tacitus, *Annals* 14.34 ; Corbulo's, *Annals* 13.40; Idistaviso, Tacitus, *Annals* 2.16; battle against the Frisii, *Annals* 4.73; against the Batavians, *Histories* 5.14; Mons Graupius, *Agric.* 29-37. Hyland sees auxiliaries as 'cannon-fodder', *Equus: the Horse in the Roman World*, 1990, 166, as does Liebeschuetz, 'The End of the Roman Army', in *War and Society in the Roman World*, J.Rich & G.Shipley (eds.), 1993, 268.

124 Bishop and Coulston, 1993, 208 on the suitability and practicality of legionary and auxiliary equipment. For a more detailed discussion of the role of auxiliaries in battle, see C.M.Gilliver, 'Mons Graupius and the role of auxiliaries in Battle', *Greece and Rome* vol.xliii, 1996, No. 1, 54-67.

125 Alterations to *pila* and spears, Plutarch, *Marius*, 25; Bishop & Coulston, *Roman Military Equipment,* 1993, 48-51; 65-66; Suetonius, *Domitian*, 10. For Legion II Parthica in Apamea, J.C.Balty & W.Van Rengen, *Apamea in Syria. The winter quarters of Legio II Parthica*, 1993.

126 Vegetius 3.9 on tribunes' responsibility for training; Trajan training his troops for Parthia, Dio 68.23. Legio III Gallica, stationed for a long period in Syria, had picked up the local habit of hailing the rising sun, Tacitus, *Histories* 3.24, though many of the legionaries were no doubt themselves Syrian. Caesar in Africa, *B.Afr.* 71.

127 On pursuit and flight, Onasander, 19; 27; 32; Vegetius 1.15; 3.14; 21-22; Frontinus, *Strat.* 2.6.

128 For the defeat by Pharnaces, *B.Alex.* 40; avoidance of battle because of proximity of marching camps, *B.Civ.* 1.82.

129 On rules governing triumphs, Valerius Maximus, 2.8.1; for Appius Claudius Pulcher, Orosius 5.4.7. Tacitus reports the triumphal insignia granted to Corbulo after a minor incursion across the Rhine in AD 47 from which he was recalled by Claudius. The governor of the other German province, Curtius Rufus, was awarded the same honour for digging a silver mine. Commenting on the cheapening of these honours, Tacitus claims that the troops who had been forced to do the work appealed to the emperor to award these honours before the governors took up office, *Annals* 11.20.

130 For casualty figures, see P.Brunt, *Italian Manpower*, 1971, Appendix 28.

131 Onasander 33. Vegetius 3.18 presents similar advice.

132 Hanson, *The Western Way of War*, 1989, 107-116 on the role of the general in the classical Greek phalanx; Plutarch, *Pyrrhus* 28, Arrian, *Anabasis* 2.10 for Alexander at Issus and 3.14, at Gaugamela leading his troops into battle; Paullus at Cannae, Polybius 3.116; for a detailed discussion of the role of the general in Roman battles, see Goldsworthy, *The Roman Army at War*, 1996, 149-162.

133 Julianus, Josephus, *B.Jud* 6.81-90; the unnamed soldiers at Cremona, Tacitus, *Hist.* 3.23; on displays of gallantry, S.Oakley, 'Single Combat and the Roman Army', *CQ* 35 (1985) 392-410 and Goldsworthy, 1996, 265-271.

134 Caesar, *B.G.*1.52, against Ariovistus, but Roman officers may have been encouraged to show initiative, as in the case of the tribune at Cynoscephalae, Livy 33.9; Pansa's injury, Appian, *B.Civ.* 3.69; on Metellus, Plutarch, *Sertorius* 21. On spreading rumours, Frontinus, *Strat.* 2.7; Onasander 23.

135 Onasander 36 on procedures after battle; Frontinus, *Strat.* 4.1 and Tacitus, *Annals* 13.36 for punishments of the cowardly; Maxfield, 1981, *The Military Decorations of the Roman Army* and Goldsworthy, 1996, 276-279 for rewards; on plunder, Shatzman, 'The Roman General's authority over booty', *Historia* xxi (1972) 177-205.

136 Philip V was criticized for his failure to bury his dead after Cynoscephalae, and they were not buried for several years, Livy 36.8.5; Plutarch, *Marius* 21 on Aquae Sextiae; on the stratagem in Spain, Frontinus, *Strat.* 2.10.1; the stratagem suggests that the extent of victory might be difficult to gauge in the aftermath of battle, and that counting the fallen was one way of discovering that Cremona: Tacitus, *Histories* 2.45; Crastinus: Appian, *B.Civ.* 2.82; Trasimene: Livy 22.7; Cannae: Livy 22.52. For full details of Greek war dead and monuments, see W.K.Pritchett, *The Greek State at War*, Part IV, 1985, and for problems in identifying the dead, P.Vaughn, 'The Identification and Retrieval of the hoplite battle-dead', in V.D.Hanson (ed.), *Hoplites: the Classical Greek Battle Experience*, 1993, 63-86. Many of her observations are as relevant for Roman warfare as for Greek.

137 Burial of the dead from the Varian disaster: Tacitus, *Annals* 1.62; commemoration at Adamklissi: *ILS* 9107; Appian, *B.Civ.* 1.43 on reactions to casualties in the Social War, and Livy 31.34 on Philip V.

138 Livy, 22.51 on the killing of the Roman wounded at Cannae; Caesar, *B.G.* 1.26 on delays after defeating the Helvetii; Dio 68.8.2 on Trajan helping the injured in Dacia.

139 Frontinus, *Strat.* 1.Preface on the use of stratagems.

140 The importance of Syracuse, Livy 29.22; plunder from New Carthage, Livy 26.47.

141 Josephus, *B.Jud.* 66.317 for the glut of gold after the capture of Jerusalem; Sallust, *B.Jug.* 97 on the consequences of the loss of Jugurtha's treasury.

142 On the supplies and siege warfare, see below; the deforestation at Jerusalem, Josephus *B.Jud.* 5.523; Caesar also notes the deforestation in the territory of Marseilles caused by his siege of that city, *B.Civ.* 2.15, and his supply problems at Dyrrachium, Caesar, *B.Civ.* 3.47-49.

143 The surrender of Noviodunum, Caesar, *B.G.*2.12; *oppida* in Britain, *B.G.*5.9; 5.21; Caesar's description of the *murus Gallicus*, *B.G.* 7.23; on the mobility of Gallic wealth, Polybius 2.18; and the campaign against the Veneti, *B.G.*3.12-14. The same factors probably added to the problems the Roman experienced fighting against German tribes.

144 Palaeopaphos, Maier & Karageorghis, *Paphos, History and Archaeology*, 1984, 192-203; Plataea, Thucydides 2.75-78; 3.52.

145 Dionysius I of Syracuse, Diodorus Siculus 14.42.1; the siege of Salamis, Diodorus 20.48, and Rhodes, 20.85-88; Alexander's assault on Halicarnassus, Arrian, *Anabasis* 1.20-23; A.Wilkins, 'Reconstructing the *Cheiroballistra*', *Journal of Roman Military Equipment Studies* 6 (1995) 5-59.

146 Lawrence, *Greek Aims in Fortification* for the defences of Syracuse; for accounts of the siege see Livy 24.34; 25.23-31, Polybius 8.3-7 and Plutarch, *Marcellus*13-20; Zonaras, 9.4 claims that Archimedes used a mirror to concentrate and aim the rays of the sun to burn up the Roman ships.

147 The siege of Adrianople, Ammianus 31.15.

148 Vitruvius, *de Architectura* 10.16.2.

149 For shelters, Apollodorus 140-144; 153-155; Vegetius 4.13-16; on the name of the tortoise, Vegetius 4.14; Caesar, *B.Civ.* 1.34-36; 56-58; 2.1-22 on Marseilles.

150 Onasander on the *testudo*, 20; Dio 49.30-1; Caesar in Britain, *B.G.* 5.9, and Josephus at Jotapata, *B.Jud.* 3.278.

151 On dimensions of towers, Josephus, *B.Jud.* 3.284 for Jotapata, 7.308 for Masada; for Aquileia, Ammianus 21.12. Apollodorus' advice, 173-174.

152 Vegetius 4.18-20 on defences against towers; on the dangers of sorties, Onasander 40; Appian, *Mithridatic War* 31 for Athens; Caesar, *B.Civ.*2.14 on Marseilles; Ammianus 19.5 on Amida.

153 Vetera, Tacitus, *Histories* 4.23 & 30. The historian's comment agrees with other Roman views on barbarian inability at this type of warfare.

154 Descriptions of rams may be found in Vitruvius, *de Architectura* 10.15; Vegetius 4.14; Josephus, *B.Jud.* 3.214-221 and Ammianus 23.4.8-9. Apollodorus' description and highly practical suggestions about the use of rams are in his treatise on siege engines 153-164.

155 Victor, *B.Jud.* 5.299; Singara and Bezabde, Ammianus 20.6 & 7.

156 Syracuse, Livy 24.34; Haliartus, Livy 42.63, and Ambracia, Livy 38.5; sacks of chaff at Jotapata, *B.Jud.* 3.222; for the siege of Marseilles, Vitruvius, *de Architectura* 10.12.

157 Caesar, *B.G.* 1.32; Cicero, *de Officiis* 1.35; Josephus, *B.Jud.* 5.277.

158 Apollodorus 143-152 on mining and firing techniques; Vegetius 4.5 for methods of countering them; Vitruvius' account of Marseilles, *de Arch.*10.16.11-12.

159 Aeneas Tacitus on detecting mines, 37; Vitruvius also describes the use of such a device by the Greek defenders of Apollonia when it was attacked by Philip V of Macedon in 214 BC, *de*

Architectura 10.16.9; Ambracia, Livy 38.7; Dura Europos, Rostovzeff et al., *The Excavations at Dura Europos*, 1929-52.

160 Athens, Appian, *Mithridatic War.* 37; Jerusalem, Josephus, *B.Jud.* 6.33-67.

161 New Carthage, Livy 26.48; Sallust, *Catiline* 7; for details of the *corona muralis* and other military decorations, see Maxfield, *The Military Decorations of the Roman Army*, 1981.

162 E.W. Marsden, *Greek and Roman Artillery: Historical Development*, 1969, and *Greek and Roman Artillery: Technical Treatises*, 1971.

163 Marcellus' artillery on ships at Syracuse, Plutarch, *Marcellus* 14; Josephus, *B.Jud.* 3.257 for Josephus' often exaggerated claims about Roman artillery. On the sound of artillery, Ammianus 19.6 and Josephus, *B.Jud.* 3.247.

164 For inscribed slingshot, *E.E.* VI.1885; Appian, *B.Civ.* I.47 for the siege of Athens. A.Bosman has postulated the direction of attacks on the Roman fort of Velsen in AD 28 through finds of lead shot which had been fired by the defenders, 'Pouring lead in the pouring rain', *JRMES* 6 (1995) 99-103.

165 Clay slingshot, Caesar, *B.G.*4.8; on naphtha, Dio on Hatra, 76.10. Lucullus' siege engines in 69 BC were all destroyed by the Parthians pouring burning naphtha on them, Dio 36.1.

166 Vegetius 4.18; Ammianus 23.4.14. Livy describes a *pilum*-like incendiary javelin used by the Spanish against Hannibal at Saguntum as a *falarica*, 21.8.10. The use of the word may have changed over time, but this illustrates the difficulties in using technical vocabulary.

167 Dio l.34.2 for Actium; for Hannibal's ruse, Cornelius Nepos, *Hannibal* 11.5; and for Hatra, Herodian 3.9.5.

168 On encouraging the soldiers, Onasander 42; and on Titus' concerns for speed at Jerusalem, Josephus, *B.Jud.* 5.502-7; for the advice on sudden assaults, Onasander 39; 42; and Vegetius 4.12.

169 Caesar, *B.Civ.* 3.80-81 on the capture of Gomphi.

170 Apollodorus 140 and Vegetius 4.28 on circumvallations; for the use of pigeons, Frontinus, *Strat.*3.13.8.

171 Josehpus *B.Jud.* 5.491-511; 6.12.

172 Vegetius 4.7-11 and Vitruvius 5.9.8-9 on supplies and stockpiling.

173 Mutina, Appian, *B.Civ.*3.49; preparations in Judaea, Josephus, *B.Jud.*2.573-76; surrender of Uscana, Livy 43.18; and Paetus' surrender, Tacitus, *Annals* 15.15.

174 Water shortages at Jerusalem, Dio 65.4

175 Caesar, *B.Civ.*3.80-81 on the capture of Gomphi.

176 Uxellodunum, Hirtius, *B.G.*8.40-43; Sertorius' attacks on Metellus' supply lines and the siege of Langobrigae, Plutarch, *Sertorius* 12-13.

177 Frontinus, *Strat.* 3.4 on preventing supplies arriving; 3.14 on getting supplies through a blockade; 3.15 on giving the impression of being well supplied; Jerusalem, *B.Jud.*5.521. A variation on this is reported by Caesar: at Dyrrachium his soldiers were forced to make loaves out of roots. When the Pompeians taunted them about their starving state, they threw the loaves at them, and the Pompeians became worried about the sort of men they were facing, who were prepared to eat such stuff, *B.Civ.* 3.48.

178 Alesia, Caesar, *B.G.* 7.78; Cremna, Zosimus 1.69-70.

179 New Carthage, Frontinus, *Strat.* 3.9.1; Marius, Sallust *B.Jug.* 92-94; Frontinus *Strat.* 3.9.3; the geese on the Capitol, Livy 5.49; Frontinus' section on this topic, *Strat.* 3.2; though not included by Frontinus, the Trojan Horse is a stratagem of exactly this type.

180 Sulla, Frontinus, *Strat.* 2.9.3 Appian, *B.Civ.* 1.92-4; Corbulo, Frontinus, *Strat.* 2.9.4; Machaerus, Josephus, *B.Jud.* 7.200-206.

181 Requests by 'barbarian' armies for Romans to surrender fortified sites: Caesar, *B.G.* 5.38; Tacitus, *Histories* 4.21-4; Ammianus 31.15. Titus' request at Jerusalem, *B.Jud.* 5.360.

182 On the rules of Medieval warfare, Keen, *The Laws of War in the Middle Ages*, 1965, chapter 8, and Bradbury, *The Medieval Siege*, 1992, chapter 10. Caesar discussing terms with the Gauls at Alesia, *B.G.* 7.89-90; requests to surrender: Noviodunum, *B.G.* 2.12; Atuatuci, *B.G.* 2.32; Sontiates, *B.G.* 3.21; Vellaunodunum, *B.G.* 7.11. Hatra, Cassius Dio 76.12. Reman and Busan, Ammianus 18.10; Ammianus reports Sapor's confidence that Amida would surrender, and surprise when his approach was met by a hail of missiles at 19.1

183 For the capture of Phocaea, Livy 37.32; Capsa, Sallust, *B.Jug.* 91; and Locha, Appian, *Punica* 15.

184 For details, see G.M.Paul, *A Historical Commentary on Sallust's Bellum Jugurthinum*, ARCA Classical and Medieval Texts, Papers & Monographs 13, 1984, 226-7. For a more general

discussion of the rules of war, see Gilliver, 'The Roman Army and Morality in War' in ed. Lloyd, *Battle in Antiquity*, 1996, 219-238.

185 Josephus, *B.Jud.* 6.284; 403-408.

186 The treatment of New Carthage, Polybius 10.15.4-16.

187 Caesar, *B.Civ.* 2.22.

188 The sack of Avaricum, Caesar *B.G.*7.28; Carthage, Appian, *Punica* 118; on the lack of military discipline in civil war, Tacitus, *Histories* 2.29; 3.7, on the sack of Cremona, *Histories* 3.33, and on Capua, Livy 26.16.

189 Josephus, *B.Jud.* 7.259. Suicide was not unusual in the end stages of a siege; Josephus himself had escaped a suicide pact at the fall of Jotapata, *B.Jud.* 3.386-91.

190 On military service as a source of enrichment, Plautus, *Bacchides* 1075; Sallust *B.Jug.*68; for a general discussion of the elite and enrichment, Harris, *War and Imperialism in Republican Rome*, 1985.

191 Polybius' description of the plundering of New Carthage at 10.15 is discussed by A.Ziolkowski, '*Urbs Direpta*, or how the Romans sacked cities', in ed. Rich & Shipley, *War and Society in the Roman World* 1995, 69-91. Polybius' figures for the plunder taken at New Carthage differ to Livy's and it is not possible to determine whose are the more accurate. In the imperial marching camp described by Pseudo-Hyginus captured booty was kept in the *quaestorium* towards the rear of the camp, even though a quaestor no longer attended the army, *Ps.Hyginus* 18; refugees at Jerusalem, Josephus, *B.Jud.* 5.550-52.

Appendix 1: Roman military treatises

Cato the Elder, *de Re Militari* (on military matters), or *de Disciplina Militari* (on military training and discipline), as cited by Vegetius (1.15). Cato was possibly the first Roman to compile a military treatise. Only fifteen fragments survive quoted in other works but the treatise probably consisted of only one book, and included information on unit organization, methods of maintaining discipline, march and battle formations, the use of specialist troops and the taking of auspices. Frontinus quotes Cato (*Strat.* 1.6.10), and Vegetius lists him as one of his sources (1.8). Much of Vegetius' information on legionary organization and deployment may derive from this treatise.
Not extant.

Asclepiodotos, Τέχνή Τακτίκή (Tactics). A Greek philosopher and pupil of Posidonius writing in the first century BC. This is the earliest surviving complete military treatise dating to the Roman period. The work, possibly derived from an earlier composition by Posidonius, is a detailed description of the Greek phalanx and its tactics, including sections on the disposition of light troops and cavalry, and the use of chariots and elephants in warfare.
Text and translation: The Illinois Greek Club, *Aeneas Tacticus, Ascleptiodous, Onasander*, Loeb Classical Library, 1923.

Cincius Alimentus, *de Re Militari*. The work, cited briefly by Aulus Gellius, is probably that of the constitutional antiquarian writing at about the time of Augustus. The treatise, which does not survive, was at least six books long, with the third including the declaration of war, the fifth the levying of troops, and the sixth the organization of units. Sections from the treatise are quoted by Aulus Gellius (16.4).
Not extant. Text and translation of the sections quoted by Aulus Gellius: J.Rolfe, Aulus Gellius, *Attic Nights*, Loeb Classical Library, 1927-8.

Vitruvius, *de Architectura*. M.Vitruvius Pollio was an architect working under Augustus. During the civil wars he had served Caesar and then Octavian as a military engineer and worked with artillery in particular (*de Arch.* 1 pref. 2). He may have been one of Caesar's engineers at the siege of Marseilles in 49 BC. His treatise is a general one on architecture, but the tenth book is on mechanics and includes detailed descriptions of various artillery pieces.
Text and translation: H.Granger, Vitruvius, *The Ten Books on Architecture*, Loeb Classical Library, 1934.

Athenaeus Mechanicus, περί μήχάνήματων (on military machines). Athenaeus was probably a contemporary of Vitruvius, and his treatise is very similar to Vitruvius' section on military engines. Both Athenaeus and Vitruvius probably based their works on that of Aegesistratus whom they acknowledge as a source.
Text and German translation: R.Schneider, 'Griechische Poliorketiker', *Abhandlungen der Köninglichen Geschellschaft der Wissenschaften zu Göttingen*, 12, 1912.

Cornelius Celsus, Title unknown, but the author is cited as an authority by Vegetius (1.8). Celsus wrote an encyclopaedia at the time of Tiberius of which the military treatise was a part. Only fragments of his *de compositione medicamentorum* survive. The military section may have been a fairly general one similar to that produced by Cato, and possibly based on it.
Not extant.

Pliny the Elder, *de iaculatione equestri* (on throwing the javelin from horseback) was written whilst Pliny was prefect of a cavalry unit based in Germany, probably under Claudius (Pliny, *Letters* 3.5). Although the work is lost, Pliny the Elder quotes it briefly in his encyclopaedic Natural History (*NH* 8.159, 162). The single book included advice on the best type of cavalry horse, and ways in which a trained horse could assist the rider in battle.
Not extant.

Onasander, Στρατήγίκός (on generalship). This Greek philosopher's treatise on the art of generalship was dedicated to a Quintus Veranius, probably the same man who was consul in AD 49 and who died whilst governor of Britain in c.AD 58. The treatise is rather different to other surviving works as it lays great emphasis on the moral qualities and other abilities which the author considered a general should possess. Onasander then gives advice on how the commander should conduct himself and proceed in a variety of situations, often giving fairly general guidelines rather than laying down strict instructions for each situation as Vegetius does.
Text and translation: The Illinois Greek Club, *Aeneas Tacticus, Ascleptiodous, Onasander*, Loeb Classical Library, 1923.

Frontinus, *Strategemata* (Stratagems). This is one of the very few military works in Latin to have survived. The author, Sextus Julius Frontinus, held the consulship three times, was governor of Britain under Vespasian, and later *curator aquarum*, in charge of Rome's water supply. He seems to have been a keen author of handbooks relating to the various offices he held, producing works on the art of war (*de scientia militari*, or *de officio militari*, according to the 6th century antiquarian John Lydus), on the aqueducts of Rome (*de Aquis*), and possibly also surveying. The *de Aquis* was written to help him and his successors understand the administration of Rome's water supply (*de Aquis* pref. 2), and his military works, unlike those of the Greek philosophers, were written by an expert.

Frontinus claimed to be the only man interested in military science to have reduced its rules to a system, and he considered his *strategemata* to be completing the task begun by his treatise on warfare. According to Vegetius, Trajan thought very highly of the treatise, and it may have included information on siege machinery as well as on strategy. However, Frontinus did not include siege machinery in the *strategemata* because he believed that the development of machines and engines for siege warfare had long since reached its limit (*Strat.* 3 pref.). The *strategemata* may have been intended as an appendix for the general treatise, but also belongs to a different literary genre of collections of examples.

The *strategemata* comprises four books with over 400 examples of military stratagems, mostly taken from historians and referring to the classical Greek, Hellenistic and Roman periods. Each book covers a particular aspect of campaigning; preparations for battle (book 1); battles, ambushes and retreats (book 2); sieges (book 3); general topics including discipline, justice and sayings (book 4). There is some dispute about the authorship of this final book which may have been written by Frontinus, or added later.
Text: Ireland, *Iul. Frontini Strategemata*, Leipzig 1990.
Text and translation: C.E.Bennett, *The Stratagems and the Aqueducts of Rome*, Loeb Classical Library 1950.

Pseudo-Hyginus, *de munitionibus castrorum* or *de metatione castrorum* (on fortifiying, or surveying, a camp). Author and date are disputed. The work is ascribed to Hyginus Gromaticus because the text survives as part of a manuscript of treatises on land surveying. It was probably written by a military surveyor in the late first or early second century AD. This is a detailed work explaining a new method of organization for temporary camps, and includes some information on camp location and defences.
Text and translation: C.M.Gilliver, 'The *de munitionibus castrorum*: Text and Translation.' (Teubner text of Grillone), *JRMES* 4 (1993).
Text, French translation & commentary: M.Lenoir, *Pseudo-Hygin, des Fortifications du Camp*, Association G.Budé, Paris 1979.

Aelian, Τακτίκή θεωρία (Theory of tactics). Aelian was a Greek philosopher and contemporary of Frontinus. He dedicated his work to the emperor, probably Trajan. Aelian's treatise, like that of Asclepiodotos, was a description of the workings of the Macedonian phalanx.
Text and German translation: H. Köchly & W.Rustow, *Griechische Kriegsschriftsteller*, 1855.

Heron of Alexandria, βελόπόϊϊκα (Belopoeica); χείρόβαλλίστρα (Cheiroballistra). A Greek artillery technician writing in the late first or early second century AD. The first of these works is based on a third century BC treatise written by Ctesibius and therefore reflects earlier usages, but the second includes recent developments.
Text and translation of Cheiroballistra: E.W.Marsden, *Greek and Roman Artillery, Technical Treatises*, chapter 6, Oxford 1971, repr. 1991.

Apollodorus of Damascus, πόλίόρκετίκά (on siege warfare). An architect working during the principates of Trajan and Hadrian, Apollodorus was the designer of Trajan's bridge over the Danube and Trajan's Forum in Rome. He was exiled in AD 129 and later executed by Hadrian, allegedly for criticizing the emperor's designs for the Temple of

Venus and Rome. He wrote a book on siege machinery dedicated to Hadrian which describes a variety of equipment needed for assaulting a stronghold. He does not, however, deal with the defence of strongholds.
Text and Italian translation: *l'Arte dell' Assedio di Apollodoro di Damasco*, ed. A. La Regina, Milan, 1999

Arrian, Τέχνή Τακτίκή (Tactics); Ἔκταξίς κατ' 'Αλανῶν (Order of March against the Alans, also known as *acies contra Alanos* or *ektaxis*). Arrian was a Greek senator who was governor of Cappadocia under Hadrian. He wrote three military works, of which the two mentioned above survive. The first of these, the Tactics, was dedicated to Hadrian in AD 137 and was a treatise primarily about the Hellenistic phalanx, similar to those of Asclepiodotos and Aelian above. Arrian made some attempt to make his treatise more up to date by including references to contemporary practices, such as the British chariotry. The last section of the work is a description of the *hippica gymnasia*, exercises carried out by the Roman cavalry.

The *ektaxis* probably dates to c.AD 132 when Arrian was governor of Cappadocia. It describes his proposed order of march, battle line and battle plans for dealing with a threatened Alan invasion. This is the only document of its kind dating to the early Empire.
Text: Ed. A.G.Roos, *Arrian, Scripta Minora*, Leipzig, 1967.
A translation of the *ektaxis* is provided as Appendix 2.

Polyaenus, Στρατήγήματα (*Strategica*, or Stratagems). This collection of stratagems was written between c.AD 161 and 166 by a Greek philosopher, and dedicated to the emperors Marcus Aurelius and Lucius Verus. It is similar to Frontinus' work, containing eight books of *exempla*, arranged according to individuals and including the exploits of gods and mythical heroes. The tenth century Byzantine encyclopaedia, the *Suda*, suggests that Polyaenus also wrote a *Tactica* in three books which probably dates to after the *Strategica*, but about which nothing is known.
Text and translation: P.Krentz & E.L.Wheeler, *Polyaenus Stratagems of War*, Chicago, 1994.

Tarruntenus Paternus, *de Re Militari* (on military matters). The author was a legal expert and probably the Praetorian Prefect under Commodus who was executed for treason (Dio 72.5), described by Vegetius as being the 'keenest advocate of military law' (Vegetius 1.8). The treatise dated to the time of Marcus Aurelius or Commodus. Although it no longer survives, it may have been heavily based on the *Constitutiones* of Hadrian, and it is possible that Vegetius used it when compiling his second book on the organization of the legion. Little is known of the contents though, except that it illustrated the author's knowledge of the law; his definition of *immunes* (those exempt from fatigues) in the army and list of examples was included in Justinian's *Digest* (50.6.7).
Not extant.

Emperor Julian (?), Μήχανικόί (Mechanics, or On siege machinery). The author of this lost treatise is disputed. However, from Ammianus' account of Julian's campaigns, it is clear that the emperor was very interested in siege machinery, even building a helepolis during the siege of Pirisabora in AD 363, and there is no reason to dismiss the possibility that the emperor did indeed write such a treatise.
Not extant.

Vegetius, *Epitoma rei militaris* (Epitome of military science). This was probably written in the late fourth century AD and is the only general manual on Roman military institutions to have survived. Vegetius made extensive use of earlier writers whom he lists at one point. The treatise is divided into four books: on recruitment and training; on the organization of the legion; on the order of march and of battle; on siege and naval warfare. Vegetius can be problematic as a source because it can be difficult to determine to which period the information he is using belongs. The *Epitoma rei militaris* is the only Roman military treatise to have been studied in detail, and it had an enormous influence on later treatises, including Machiavelli's *Arte della Guerra*.
Text: ed. C.Lang, *Epitoma Rei Militaris*, Lepizig, 1885.
Translations: The first three books were translated and first published by Lieutenant John Clarke of the Royal Navy, in 1767. This translation was reprinted by T.R.Phillips as 'The Military Institutions of the Romans' in *The Roots of Strategy*, in 1944. Two English translations have been published in the last few years, by L.F.Stelten, with text, *Epitoma Rei Militaris*, New York, 1990, and by N.P.Millner, translation only, with brief notes, *Vegetius' Epitome of Military Science*, Liverpool, 2nd ed. 1996.

Appendix 2: Arrian's *Order of March against the Alans*

(1) The whole army will be led by the mounted scouts, arranged in two sections under their own commander. After these will come the mounted Petraean archers also in two sections, under the command of their decurions. Next in the formation will be the men from the Ala Auriana, and stationed with them the men of the fourth cohort of Raetians, commanded by Daphnis the Corinthian. After them should be the men from the Ala of Colonists; and with them in the formation will be the Ituraeans and Cyrenaicans, and the men from the first cohort of Raetians. All these will be commanded by Demetrius. (2) Next will come the German cavalry, also in two sections, and commanded by the centurion in charge of the camp. (3) The infantry will be stationed behind these units, carrying their standards before them, the Italians and those of the Cyrenaicans who are present. Pulcher, the Prefect of the Italians, will command all of these. The Bosporan infantry will follow them under the command of Lamprocles, and then the Numidians commanded by their own prefect, Verus.

(4) They will march in ranks four abreast, and will follow their archers. Their own cavalry will protect both flanks of the marching column. Next will come the *equites singulares* and the legionary cavalry, (5) then the artillery train, then the standard of the 15th Legion and Valens the commander of the legion, his second in command, the military tribunes assigned to the expedition, and the (five) centurions commanding the first cohort of the legion. The javelinmen will be in front of the infantry's standard. The infantry will march in ranks four abreast. (6) The standard of the 12th Legion, its tribunes and centurions, will be stationed after the 15th Legion. This legion too will also march in ranks four abreast.

(7) The allies will come after the heavy infantry, first the heavy infantry from Armenia Minor and Trapezus, and the spearmen from Colchis and Rhizus. After them will come the Apulian infantry. All the allies are commanded by Secundinus, the commander of the Apulians. (8) Next will come the baggage train. The Ala of Dacians under its own commander will bring up the rear. (9) Centurions who have been posted for this purpose will keep the ranks of infantry in order. For protection, the Ala of Gauls will ride in single file on both sides of the marching column, and also the Italian cavalry. Their prefect should inspect the flanks regularly. (10) The commander of the whole army, Xenophon (Arrian) will be stationed for the most part in front of the infantry standards, but will also regularly ride up and down the whole column and inspect it to make sure they are marching in order, to order back into line those who have become disordered, and to praise those marching in proper formation. (11) This should be the order of march.

When the army has come to the appointed place, all the cavalry will wheel round in a circular manoeuvre to a square formation, and the mounted scouts will be sent ahead to high ground to look out for the enemy. Then at a given signal the men will arm in silence and when they have armed, they will dress their ranks. (12) The order of battle will be as follows: each wing of the infantry should hold the high ground, since that is their formation in such terrain. The Armenians will be drawn up on the right wing under Vasaces and Arbelus, holding the highest part of the wing because they are all archers. (13) The cohort of Italian infantry will be stationed in front of these. Plucher, the commander of the Italian cohort, will be in overall command. Vasaces and Arbelus will support him with their cavalry and infantry.

(14) The allies from Armenia Minor, the light-armed troops from Trapezus and the Rhizionian spearmen will be stationed on the left and should hold the highest part of the left wing. The 200 Apulians and the 100 Cyrenaicans should be drawn up in front of them so that the heavy infantry can protect the spearmen throwing over their heads from the higher ground. (15) In the space between the hills, the infantry of the 15th Legion will occupy the whole of the right side as well as the centre because of its greater strength. The 12th Legion will fill the remaining space on the left, as far as the extreme left of the wing. They will be drawn up in ranks eight deep in close order. (16) The first four ranks will be equipped with pikes with long slender iron heads. The men in the front rank will hold their pikes ready so that if the enemy approach them they can thrust the iron head of the pikes at the horses, particularly at their chests. (17) The second (?), third and fourth ranks should be ready to thrust with their pikes and wound the horses wherever possible and kill the riders. When the pike has pierced the shield or protective armour, because of the softness of the iron, it will bend, making the rider ineffective. (18) Behind them the next ranks will be of spearmen. In the ninth rank, behind these, shall be the foot archers: Numidians, Cyrenaicans, Bosporans and Ituraeans. (19) Artillery will be stationed on either wing to fire on the approaching enemy at extreme range from behind the entire battle line.

(20) All the cavalry will be drawn up in eight units and companies on the wings beside the infantry. One of these will be stationed on either wing behind the heavy infantry and archers; the remaining six will be stationed behind the middle of the legions... (21) The mounted archers will be stationed near to the legions so that they can shoot over them. Spearmen, javelinmen, swordsmen and axe-men will keep watch on the flanks for the signal. (22) The *equites singulares* will be stationed around Xenophon (Arrian), as well as up to 200 infantrymen from the legions as a personal bodyguard, centurions commanding the picked troops and bodyguard, and the decurions of the *equites singulares*. (23) About 100 of the light-armed javelinmen will be stationed around him so that Xenophon can regularly inspect the battle line and wherever he learns there is a weakness he can go and deal with it. (24) Valens, the legate of the 15th Legion, will command the whole of the right wing including the cavalry; the tribunes of the 12th Legion will command the left.

(25) When the troops have been drawn up in this way, they should keep silence until the enemy have approached to within firing range. When the enemy are within range they should all shout the war-cry as loudly and fiercely as possible. At the same time bolts and stones will be fired by the artillery, arrows from bows, the javelinmen will hurl their missiles, both the light-armed men and those with shields. Stones will also be thrown at the enemy by the allied troops on the tops of the hills. Since the whole assault will be from all angles and as heavy as possible, it will cause confusion amongst the horses and bring disaster to the enemy cavalry. (26) The hope is that under such a hail of missiles the Scythians (Alans) will not dare to approach the legionary formation. If they do advance, however, then the first three ranks should link up their shields together and stand shoulder to shoulder and set themselves in close formation to resist the attack as strongly as possible. The fourth rank (should hold their pikes up so that they may kill any enemy cavalrymen, but should not hold them straight up so as to make the spearmen behind overshoot with their spears). The front rank should thrust or hurl their javelins unstintingly at both the horses and their riders. (27) When the enemy has been repulsed, if it is clear that they are fleeing, the infantry will open up their ranks for the cavalry to advance, though not all the companies, but only half of them. (28) The other half will follow those in front, but more deliberately and in ranks, so that if the rout becomes a full one, those in the front of the pursuit can be relieved by fresh cavalry, and if the enemy wheel about they can be attacked as they turn. (29) At the same time the Armenian archers should advance and fire, to prevent the fleeing enemy from turning back again, and the light-armed spearmen should follow at the run. The infantry formation will not remain on the battlefield but will advance at a quick march so that if stiff resistance from the enemy should be encountered, they can again become a protective screen in front of the cavalry.

(30) This should be carried out if the enemy turns to flight immediately after the first assault. If, however, the enemy wheels about and attempts to outflank the wing, the light-armed archers are to extend the wings out to the hills. I do not think that the enemy, on seeing that the wings have been weakened because they have been extended, would thrust their way through them and break through the infantry. (31) But if they are victorious on one wing or the other, it is inevitable that their cavalry and their javelins will be exposed at an oblique angle to us. Then our cavalry should attack them, not with javelins but with swords and axes. For the Scythians, being unprotected by armour, and with their horses unprotected...

Glossary

aerarium	Military treasury.
agmen quadratum	A compact marching formation or a squared column for use in hostile territory or when about to be attacked.
antesignani	Technically, the front-line troops, the bravest men who fought in advance of the standards. Caesar sometimes used his *antesignani* for special duties.
auxilia	In the Republic, *auxilia* were troops provided by non-Italian tribes to Rome under treaty or by private arrangement between Roman general and tribal leaders. These troops were regularized during the late Republic and early Empire, to form the permanent auxiliary units of the Imperial army.
bracchium	A fortified arm running from an encampment to an objective, often a water supply, built to protect from sudden attack those moving around.
clavicula	Gateway design for a temporary camp, consisting of a curved extension of the rampart either internally or externally, to impede an enemy attacker and expose his unshielded right side to the defender.
cohors equitata	An auxiliary unit comprising part infantry and part cavalry, also known as a part-mounted unit.
cohort	A sub-unit of the late Republican and imperial legion, containing usually c.480 men. Also a unit of auxiliary troops of about the same size, or as a *milliary* unit, c.800-1000 strong.
Consul	Chief magistrate of the Roman state. The two consuls each year in the Republic commanded the army. Ex-consuls governed provinces and might be given special military commands during both late Republic and Empire.
contubernium	A 'mess unit' of the Roman army; it contained about 8 men who shared a tent or a barrack room in a permanent fort.
duplex acies	Battle formation of two lines of cohorts, usually in a 5-5 formation.
equites legionis	Cavalry unit attached to each legion, 200-300 in the mid Republic, c.120 in the Empire.
equites singulares	The cavalry bodyguard of a provincial governor, or general on campaign.
falarica	Incendiary bolt fired from a catapult.
hastati	Front line of heavy infantry in the mid Republican legion, armed with javelins (*pila*), sword, large shield and body armour.

imperator	In the Republic, a successful general was proclaimed *imperator* by his soldiers, one of the qualifications for application for a triumph. In the Empire, one of the emperor's titles; no others could be proclaimed *imperator* and only members of the imperial family triumphed.
maniple	Sub-unit of the middle Republican legion, containing two centuries.
milliary	Denotes a unit nominally 1000 strong, but usually smaller.
oppidum	Hill fort or fortified hill settlement.
pedites singulares	Infantry bodyguard of a provincial governor or general on campaign
porta praetoria	The principal gate of a campaign camp or fort, facing the *praetorium* or Headquarters tent or building.
praetentura	The front area of a campaign camp.
Praetor	Roman magistrate with *imperium*, the right to command an army. In the Empire, ex-praetors commanded legions and some smaller provinces.
praetorium	The general's tent in a campaign camp, or house of the unit commander in a permanent fort.
principes	Second line of heavy infantry in the mid Republican legion, armed as the *hastati*.
principia	Headquarters tent or building in a campaign camp or permanent fort.
Quaestor	Roman magistrate with financial responsibilities. On campaign, the quaestor was responsible for cataloguing booty, but might command troops as a subordinate.
quingenary	Denotes a unit nominally 500 strong, but usually smaller.
socii	The Italian allies who provided troops for Rome's campaigns under treaty, until the Social War of 91-89 BC.
testudo	Compact defensive formation of soldiers with shields held on all sides and above heads. Used against enemy archers, and to attack enemy strongholds. Also the name of the shelter that held the battering ram.
titulum	Gateway design for a temporary camp, consisting of a short section of rampart and ditch in front of the gate opening to prevent an enemy from charging in.
triarii	Third line of heavy infantry in the mid Republican legion, armed as *hastati*, but with spears (*hastae*) instead of *pila*.
triplex acies	Battle formation of three lines of cohorts, usually in a 4-3-3 formation.
turma	A sub-unit of Roman cavalry, containing c.30 horsemen.
velites	The light-armed troops of the mid Republican legion who acted as skirmishers and pursuers.
vexillation	A detachment of a unit or army.

Bibliography

Adams, C.E., 1995, 'Supplying the Roman Army: O.Petr.245', ZPE 109, 119-124.

Adcock, F.E., 1940, *The Roman Art of War under the Republic*, Cambridge, Mass.

Alföldi, G., 1968, *Die Hilfstruppen der Römischen Provinz Germania Inferior*, Dusseldorf.

Astin, A.E., 1978, *Cato the Censor*, Oxford.

Austin, N.J.E., 1979, 'Ammianus on Warfare', *Coll. Latomus* 165.

Austin & Rankov, B., 1995, *Exploratio*, London.

Balty, J.C. & Van Rengen, W. *Apamea in Syria. The winter quarters of Legio II Parthica*, 1993.

Bandy, A.C., 1983, *On Powers or The Magistracies of the Roman State*, American Philological Society Memoirs vol 149, Philadelphia.

Barnes, T.D., 1979, 'The date of Vegetius', *Phoenix* 33, 254-7.

Becatti, G., 1957, *Colonna di Marco Aurelio*, Milan.

Bell, M.J.V., 1965, 'Tactical Reform in the Roman Republican Army', *Historia* 14, 404-422.

Bennett, J., 1982, 'The Great Chesters "Pilum Murale"', *AA*[5] x, 200-205.

Bennett, J., 1986 'Fort sizes as a Guide to Garrison Type: A preliminary study of selected forts in the European provinces', in Unz C. (ed.) *Studien zu den Militärgrenzen Roms III*, Stuttgart.

Bingham, J., 1631, *The Art of embattling an army, or the second part of Aelian's Tactics*, London.

Birley, E., 1953, 'The Epigraphy of the Roman Army', in *Actes du Deuxième Congrés International d'Épigraphe Grecque et latine*, 226-38.

Birley, E., 1966, 'Alae and cohortes milliariae', in *corollae memoriae Erich Swobodae dedicata*, 54-67.

Birley, E., 1969, 'Septimius Severus and the Roman Army', *ES* 8, 63-82.

Birley, E., 1981, 'A note on Hyginus and the first cohort', *Britannia* 11, 51-60.

Birley, E., 1982, 'The Dating and Character of the text *de munitionibus castrorum*', in Wirth G., (ed.) *Romanitas-Christianitas. Untersuchungen zue Geschichte und Literatur der römischen Kaiserzeit. Johannes Straub zum 70*, 277-281.

Bosman, A., 1995, 'Pouring lead in the pouring rain', JRMES 6, 99-103.

Bosworth, A.B., 1977, 'Arrian and the Alani', *HSCP* 81, 217-55.

Bowman, A.K.& Thomas, J.D., 1991, 'A Military Strength Report from Vindolanda', *JRS* 81, 62-73.

Breeze, D., 1969, 'The Organization of the Legion: the First Cohort and the *Equites Legionis*', *JRS* 59, 50-55.

Breeze, D., 1980, 'Agricola the Builder', *Scottish Archaeological Forum* 12, 12-24.

Brilliant, R., 1967, 'The Arch of Severus in the Roman Forum', *MAAR* 29.

Brunt, P.A., 1971, *Italian Manpower*, Oxford.

Cagnat, R., 1913, *l'Armée romaine d'Afrique et l'occupation militaire l'Afrique sous les Empereurs*, Paris.
Campbell, D.B., 1987, 'Teach yourself how to be a general', *JRS* 77, 13-29.
Campbell, D.B, 1994, *The Roman Army: a Sourcebook*, London.
Chantraine H. (ed.), 1984, *Das römische Neuss*, Stuttgart.
Cheesman, G.L., 1911, 'Numantia', *JRS* 1, 180-6.
Cheesman, G.L., 1914, *The Auxilia of the Roman Imperial Army*, Oxford.
Cichorius, C., 1896, *Die Reliefs der Traianssäule*, Berlin.
Collingwood, R.G., and Richmond, I.A., 1969 *The Archaeology of Roman Britain*, London.
Connolly, P., 1981, *Greece and Rome at War*, London.
Cooper, P.K., 1968, *The Third Century Origins of the 'New' Roman Army*, unpublished University of Oxford D.Phil. thesis.
Crawford, O.G.S., 1949, *The Topography of Roman Scotland*, Cambridge.
Crump, G.A., 1975, 'Ammianus Marcellinus as a military historian', *Historia* 27, 508-510.
Curle, J., 1911, *A Roman Frontier Post and its People: the Fort of Newstead in the Parish of Melrose*, Glasgow.
Dain, A., 1967, *Les Stratégistes Byzantines*, Paris.
Davies, R.W., 1968, 'Roman Wales and Roman Military Practice-Camps', *Archaeologica Cambrensis* 117 103-20.
Davies, R.W., 1968, 'Fronto, Hadrian and the Roman Army', *Latomus* 27, 75-95.
Davies, R.W., 1971, 'Cohortes Equitatae', *Historia* 20, 751-63.
Davies, R.W., 1974, 'The Daily Life of the Roman Soldier under the Principate', *ANRW* II.1, 299-338.
Davison, D.P., 1989, *The Barracks of the Roman Army from the First to Third Centuries AD*, BAR Int. Series 472.
Delbrück, H., 1975, *History of the Art of War within the Framework of Political History*, English translation, Nebraska.
Dennis, G.T. (ed.), 1985, *Maurice's Strategicon*, Philadelphia.
Dennis, G.T. (ed.), 1985, *Three Byzantine Military Treatises*, CFHB 25, Washington.
Dilke, O.A., 1971, *The Roman Land Surveyors: An Introduction to the Agrimensores*, Newton Abbot.
Dillon, Lord, 1814, *The Art of Embattling an Army, being the second part of Aelian's Tactica*, London.
Von Domaszewski, A. / Dobson B., 1967, *Die Rangordnung des römischen Heeres*, 2nd ed. Köln.
Durry, M., 1938, *Les Cohortes Prétoriennes*, Paris.
Erdkamp, P., 1999, *Hunger and the Sword: warfare and food supply in Roman Republican wars (264-30 BC)*, Amsterdam.
Ferrill, A. *The Fall of the Roman Empire: the Military Explanation*, 1986, London.
Fink, R.O., 1971, *Roman Military Records on Papyrus*, Philological Monographs of the American Philological Association 26, Cleveland, Ohio.
Fraccaro, P., 1934, 'Polibio e l'accampamento romano', *Athenaeum* 12, 154-161.
Freeman, P. & Kennedy, D., (eds.), 1986, *The Defence of the Roman and Byzantine East*, BAR Int. Series 297, Oxford.

Frere, S., 1980, 'Hyginus and the First Cohort', *Britannia* 11, 51-60.

Frere, S., & St. Joseph, J.K., 1983, *Roman Britain from the Air*, Cambridge.

Frere, S., & Lepper, F., 1988, *Trajan's Column*, Gloucester.

Furhmann, M., 1960, *Das Systematische Lehrbuch*, Göttingen.

Gabba, E., 1983, 'Technologica militare antica', *Tecnologia, economia e societa*, Atti del Convengo di Como, 219-234.

Gabba, E., 1976, *Republican Rome, the army and the allies*, Oxford.

Garlan, Y., 1972, *La guerre dans l'antiquité*, Paris.

Gichon, M., 1986, 'Aspects of a Roman army in war according to the *Bellum Judaicum* of Josephus', in ed. Freeman, P. & Kennedy, D., *The Defence of the Roman and Byzantine East*, BAR Int. series 297, 287-310.

Gilliver, C.M., 1993, 'Hedgehogs, Caltrops and Palisade Stakes', *Journal of Roman Military Equipment Studies* 4, 49-54

Gilliver, C.M., 1996, 'Mons Graupius and the role of auxiliaries in Battle', *Greece and Rome* vol.43.1, 54-67.

Gilliver, C.M., 1996, 'The Roman Army and Morality in War' in ed. Lloyd, *Battle in Antiquity*, 219-238, London.

Goldsworthy, A.K, 1996, *The Roman Army at War*, Oxford.

Goodman M.D, & Holladay, A.J., 1986, 'Religious Scruples in Ancient Warfare', *CQ* 36, 151-171.

Hansen, M.H., 1993, 'The Battle Exhortation in Ancient Historiography. Fact or Fiction?' *Historia* 42.2, 161-180.

Hanson, V.D., 1989, *The Western Way of War*, Oxford.

Hanson, W.S., 1987, *Agricola and the Conquest of the North*, 123-125.

Hanson, W.S., 1978, 'Roman Camps north of the Forth-Clyde isthmus: the evidence of the termporary camps', *PSAS* 109 140-150.

Hanson, W.S., 1987, *Agricola and the Conquest of the North*, 132-3.

Harris, W.V., 1985, *War and Imperialism in Republican Rome 327-70 BC*, Oxford.

Hassall, M.W.C., 1970, 'Batavians and the Roman conquest of Britain', *Britannia* 1, 131-136.

Hassall, M.W.C., 1983, 'The Internal Planning of Roman Auxiliary Forts' in ed. B.Hartley & J.Wacher, *Rome and her Northern Provinces*, 96-131.

Hawkes, C., 1929, 'The Roman Siege of Masada', *Antiquity* 3, 195-213.

Hawkes, C., 1929, 'Review of Schulten, *Numantia'*, *JRS* 19, 99-102.

Hawkes, C., & Richmond, I.A., 1934, 'Review of Schulten, *Masada'*, *JRS* 24, 72-75.

Hazell, P.J., 1981, 'The Pedite Gladius', *Antiquaries Journal* 41, 73-82.

Holder, P.A., 1980, *The Auxilia from Augustus to Trajan*, BAR Int. Series 70, Oxford.

Holmes, T.Rice, 1911, *Caesar's Conquest of Gaul*, Oxford.

Hyland, A., 1990, *Equus: The Horse in the Roman World*, London.

James, S., 1983, 'Archaeological Evidence for Roman Incendiary Projectiles', *Saalburg Jahrbuch* 40, 142-43.

Jarrett, M.J., 1994, 'Non-legionary Troops in Roman Britain: Part One, The Units', *Britannia* 25, 35-77.

Johnson, A., 1983, *Roman Forts of the First and Second Centuries AD in Britain and the German Provinces*, London.

Johnson, S., 1983, *Late Roman Fortifications*, London.

Jones, G.D.B., 1965, 'Ystradfellte and Arosfa Gareg: Two Roman marching camps', *BBCS* 21, 174-8.

Jones, G.D.B., 1970, 'The Roman Camps at Y Pigwn', *BBCS* 23, 100-103.

Junklemann, M., 1986, *Die Legionen des Augustus. Der römische Soldät im archäologischen Experiment*, Mainz.

Kandler, M., & Wetters, H., 1986, *Der römische limes in Österreich*, Vienna.

Keay, S., 1988, *Roman Spain*, London.

Kennedy, D., 1986, '"European Soldiers" and the Severan Siege of Hatra', in Freeman, P.& Kennedy, D. (eds.), 397-409.

Keppie, L., 1984, *The Making of the Roman Army*, London.

Kiechle, F., 1964, 'Die Taktik des Flavius Arrianus', *BRGK* 45, 88-129.

King, A.C., 1990, *Roman Gaul and Germany*, London.

Kromayer, J., & Veith, G., 1922, *Schlachten-Atlas zue antiken Kriegsgeschichte*, Leipzig.

Kromayer, J., & Veith, G., 1928, *Heerwesen und Kriegführung der Griechen und Römer*, München.

Lacotse, E., 1890, 'Les poliorcétiques d'Apollodore de Damas', *REG* 3, 230-280.

Lammert, F., 1931, 'Die römische Taktik zu Beginn der Kaiserzeit und die Geschichts Schreibung', *Philologus* 23.2.

Lander, J., 1984, *Roman Stone Fortifications*, BAR Int. Series 206, Oxford.

Lawrence, A.W., 1979, *Greek Aims in Fortification*, Oxford.

Le Bohec, Y., 1989, *La Troisième Légion Auguste*, Paris.

Le Bohec, Y., 1993, *The Imperial Roman Army*, London.

Lenoir, M., 1977, 'Lager mit *Claviculae*', *MEFR* 89, 697-722.

Lenoir, M., 1979, *Pseudo-Hygin, des Fortifications du Camp*, Association G.Budé, Paris.

Liebeschuetz, W., 1993, 'The End of the Roman Army', in *War and Society in the Roman World*, J.Rich & G.Shipley (eds.), 265-276.

Luttwak, E.N., 1976, *The Grand Strategy of the Roman Empire*, Baltimore.

Marchant, D., 1990, 'Roman weapons in Great Britain: a case study: Spearheads, problems in dating and typology', *JRMES* 1, 1-6.

Marquardt, J., 1891, *De l'organisation militaire chez les romaines; Manuel des antiquités romaines*, tome xi, Paris.

Marsden, E.W., 1969, *Greek and Roman Artillery: Historical Development*, Oxford.

Marsden, E.W., 1971, *Greek and Roman Artillery: Technical Treatises*, Oxford.

Maxfield, V., 1981, *The Military Decorations of the Roman Army*, London.

Maxwell, G.S., 1980, 'Agricola's Campaigns: the evidence of the temporary camps', Scottish Archaeological Forum 12, 25-54.

Mazzarino, S., 1971, 'La Tracizione sulla Guerre tra Shabuhr I e l'Impero Romano: "Prospettive" e "Deformazione Storica"', *Acta Antiqua* 19, 59-82.

Napoléon III, 1865-6, *Histoire de Jules César, II, Guerre des Gaules*, Paris.

Nash-Williams, V.E., & Jarrett, M., 1969, *The Roman Frontier in Wales*, Cardiff.

Oakley, S., 1985, 'Single Combat and the Roman Army', *CQ* 35, 392-410

Parker, H.M.D., 1928, *The Roman Legions*, Oxford.

Parker, H.M.D., 1932, 'The *antiqua legio* of Vegetius', *CQ* 26, 137-49.

Paul, G.M. *A Historical Commentary on Sallust's Bellum Jugurthinum*, ARCA Classical and Medieval Texts, Papers & Monographs 13, 1984, 226-7.

Pitts, L.F., & St. Joseph, J.K., 1985, *Inchtuthil: The Roman Legionary Fortress*, London.

Pritchett, W.K., 1985, *The Greek State at War*, Part IV, California.

Rainbird, J.S., 1969, 'Tactics at Mons Graupius', *CR* 19, 11-12.

Rosenstein, N.S., 1990, *Imperatores Victi: Military Defeat and Aristocratic Competition in the Middle and Late Republic*, California.

Rawson, E., 1971, 'The literary sources for the pre-Marian Roman Army', *PBSR* 39, 13-31.

Richmond, I.A., & McIntyre, J., 1934, 'The Roman Camps at Rey Cross and Crackenthorpe', *TCWAAS* 34, 50-61.

Richmond, I.A., 1940, *The Romans in Redesdale*, Northumberland County History 15.

Richmond, I.A., 1955, 'Roman Britain and Roman Military Antiquities', *Proceedings of the British Academy* 41, 297-315.

Richmond, I.A., 1962, 'The Roman Siege-Works at Masada', *JRS* 52, 142-155.

Rivet, A.L.F., 1971, 'Hillforts in Action', in Hill, D., & Jesson, M. (eds.), *The Iron Age and its Hillforts*, 109-202, Southampton.

Robinson, H.R., 1975, *The Armour of Imperial Rome*, London.

Roxan, M., 1986, 'Roman Military Diplomata and Topography', in *Studien zu den Militärgrenzen Roms III*, 768-778.

Saddington, D.B., 1975, 'The development of Roman Auxiliary Forces from Augustus to Trajan', *ANRW* II.3, 176-201.

Saddington, D.B., 1982 *The Development of the Roman Auxiliary Forces from Caesar to Vespasian,* Harare.

Sander, E., 1940, 'Die antiqua ordinatio legionis des Vegetius', *Klio* 32, 382-91.

Schenk, D., 1930, 'Flavius Vegetius Renatus. die Quellen der Epitoma Rei Militaris', *Klio* Beheift 22.

Shatzman, I., 1972 'The Roman General's authority over booty', *Historia* 21, 177-205.

Schulten, A., 1914-31, *Numantia I-IV*, Munich.

Schulten, A., 1933, *Masada: Die Burg des Herodes und die römischen Lager*, Leipzig.

Settis, S. *et al.* (ed.), 1988, *La Colonna Traiana*, Torino.

Silhanek, D.K., 1971, 'Vegetius' Epitoma Books 1 & 2: a translation and commentary', New York University unpublished PhD thesis.

Spaulding, O.A., 1933, 'The Ancient Military Writers', *CJ* 28, 657-669.

Speidel, M.P., 1992a, 'The Soldier's Servants', MAVORS 8, 342-350.

Speidel, M.P., 1992b, *The Framework of an Imperial Legion*, Cardiff.

Stadter, P.A., 1978, 'The *Ars Tactica* of Arrian: Tradition and Originality', *CP* 73, 117-128.

Stadter, P.A., 1980, *Arrian of Nicomedia*, Chapel Hill.

Syme, R., 1958, *Tacitus*, Oxford.

St. Joseph, K., 1970, 'The Camps at Ardoch, Stracathro and Ythan Wells: Recent Excavations', *Britannia* 1, 163-78.

St. Joseph, J.K., 1973, 'Air Reconnaissance in Roman Britain, 1969-72', *JRS* 63, 214-246.

St. Joseph, J.K., 1977, 'Air Reconnaissance in Roman Britain, 1973-76', *JRS* 67, 125-61.

St. Joseph, J.K., 1983, 'The Camp at Durno, Aberdeenshire, and the site of Mons Graupius', *Britannia* 9, 271-87.

Roth, J.P., 1999, *The Logistics of the Roman Army at War (264 BC – AD 235)*, Leiden.
Vaughn, P., 1993 'The Identification and Retrieval of the hoplite battle-dead', in Hanson, V.D. (ed.), *Hoplites: the Classical Greek Battle Experience*, 63-86.
Walbank, F., 1957-1979, *An Historical Commentary on Polybius*, Oxford.
Watson, G.R., 1983, *The Roman Soldier*, London.
Webster, G., 1985 *The Roman Imperial Army*, 3rd ed., London.
Wellesley, K., 1975, *The Long Year, AD 69*, London.
Wescher, C. (ed.), 1867, *Poliorcétique des Grecs*, Paris.
Wheeler, E.L., 1977, 'Flavius Arrianus: A Political and Military Biography', Duke University unpublished PhD thesis.
Wheeler, E.L., 1978, 'The Occasion of Arrian's Tactica', *GRBS* 19, 351-65.
Wheeler, E.L., 1979, 'The Legion as Phalanx', *Chiron* 9, 303-18.
Wilkins, A., 1995, 'Reconstructing the *Cheiroballistra*', *Journal of Roman Military Equipment Studies* 6, 5-59.
Yadin, Y., 1966, *Masada*, London.
Ziolkowski, A., 1995, '*Urbs Direpta*, or how the Romans sacked cities', in ed. Rich & Shipley, *War and Society in the Roman World*, 69-91.

Index